ANTARCTICA BECOMES HER

Extreme latitudes series, Volume 1

A Photo Essay By
MELISSA HAEFFNER

Published by Bitingduck Press

ISBN: 9781938463761
Website: https://extremelatitudes.com

For information contact
Bitingduck Press, LLC
Altadena, California
notifications@bitingduckpress.com
http://www.bitingduckpress.com

Publisher's Cataloging-In-Publication Data
Haeffner, Melissa [1978--   ]
Antarctica Becomes Her
Color, black and white photographs, graphs/by
Melissa Haeffner—1st edition—Altadena, CA:
Biting Duck Press, 2019
190 pgs. cm
Volume I of a series: "Extreme Latitudes"
ISBN: 978-1-938463-76-1
[1. Antarctica—pictorial work—landscape
photography. 2. Women explorers.] I. Title
2019943330

Table of Contents

Introduction

In 2016, a women-only expedition traveled to Antarctica. Seventy-six women, mostly research scientists, embarked on this journey. We were selected by a leadership development program called Homeward Bound, whose mission it is to 'heighten the influence and impact of women in making decisions that shape our planet.' The intent was a year-long leadership development program and a three-week expedition in order to build a global network of women leaders.

We visited two operating research stations (Carlini, Argentina and Palmer Station, United States) as well as the abandoned Argentine/Spanish research base built on a Norwegian whaling station on Deception Island and the Port Lockroy (United Kingdom) post office. We saw penguins and icebergs, seals and seafarers. But more importantly, we saw women who are making a tremendous impact on science.

This is the story about how we became a part of Antarctica and how Antarctica became a part of us.

Most glaciological research has a history of excluding women. Take, for example, this early 1900s letter from three women to Ernest Shackleton who asked to join his expedition:

> "We are three strong healthy girls, and also gay and bright, and willing to undergo any hardships, that you yourself undergo. If our feminine garb is inconvenient, we should just love to don masculine attire. We have been reading all books and articles that have been written ondangerous expeditions by brave men to the Polar regions, and we do not see why men should have the glory, and women none, especially when there are women just as brave and capable as there are men."

Alas, Shackleton did not take the female applicants. Nor did the British Antarctic Expedition when 1,300 women applied to join in 1937. However, women have been valuable in Antarctic discoveries. Indeed, women's bodies have been instrumental. Literally, Argentina sent a pregnant woman to have the first "native Antarctican" in an attempt to make a claim for Argentine ownership. Fast forward to 2016 when nearly a third of South Pole researchers were women. Despite the great strides in including women scientists in polar sciences in general, both poles must wrestle with their historical images as the White Ladies of the Pole, Waiting to be Won (Punch, 1875 quoted in Blackadder, 2015).

When people ask about my travels, I inevitably tell them about the women I meet. In this book, you will find some of the most impressive women I met during my travels in Antarctica. To be sure, you will see penguins and icebergs too. But I want to showcase women because women weren't always welcome in Antarctica. I guess I'm fascinated by women's life stories because I only get one life, but I get a window into other paths I could have taken through theirs. In producing this book, I hope to inspire young girls to do the same - to "try on" the life of a scientist in Antarctica. Could you see yourself as a scientist in Antarctica?

This is not my first book. I went to Ghana when I was at the Massachusetts Institute of Technology. My engineering professor designed a personal water filtration system to sell in Ghana. But to my professor's credit, she noticed what others might not have - that young ladies were already selling filtered water on the street. She wondered if her product would interfere with these women's livelihoods. So she asked me to go to Ghana with her to find out what I could about these young ladies' lives. I ended up talking to 100 water sellers. Some showed me their houses and introduced me to their families. They were amazing. They were selling water on the streets to pay for high school. Their parents were proud. They were often the only ones in their families who could read and write and speak English. They told me they sold water because they didn't want to sit around watching television. They wanted to be soccer players and doctors when they grew up. Yes, they lived in some of the most impoverished and economically exploited places, but they had hope. They let me take their portraits. I wanted to show people so they could be inspired like I was.

So, when I got to Antarctica, and I saw the bursting pride and life of the women participants, I took out my camera once again. Pictures are powerful because I don't want to tell their stories. I don't need to tell their stories. Because I took the photos, they are, of course, my gaze of how I see them. After all, we had been on a ship for several weeks. You only see the raw beauty when you take away all the stuff - the make-up and hairspray - and when it doesn't matter what you are wearing because you are wearing everything you own to stay warm! The ambitiousness shows through. That's why I wanted to include a picture of each scientist - so the reader could see what I see.

We need to see women as a part of Antarctica.

We each experienced place differently I think based on who we left behind. There was one phone that cost $10 per minute. So, some people would call family and kids. There was a computer, but we had to pay $10 per half-hour for Internet. Unfortunately, the connection was so slow, we couldn't send pictures, audio, or video. We did not have the communication we were used to. Even so, each of us were constantly doing things for people back home. Several women brought stuffed animals or flags and took pictures with them for their folks back home. It was like we were traveling for more than just ourselves. So, I also wanted to create this book to share the journey with others, because that is ultimately what we were all trying to do.

Life is about the journey a much as it is about the destination.

I wanted to include the itinerary and the pictures of our faculty and crew to show how our journey was as much directed by the rough Antarctic seas as it was curated by the small army of expedition leaders. During the trip, I would visit the crew on the bridge and copy their log books - coordinates, temperature, wind direction, cloud cover. I wanted to know as much about the physical environment as about the people I encountered. Later, I interviewed our expedition leader, Greg Mortimer, who was famous for climbing Mount Everest without oxygen among other exploits. He remembered in full detail how he made each hour-by-hour decision. He has a theory about how to explore Antarctica. If you experience the harshness when you first jump right in, he says, you will be too scared to enjoy the rest. It's better to start off with the beauty and then find adventure.

We think of Antarctica as a pristine, untouched place, but there is more human impact on Antarctica than you would expect. At the same time, we don't always have control over how we experience Antarctica. There were times when the ice locked us in and we had to change our plans. That is why I wanted to add the itinerary, not as a deep reflection but a focus on place and how we were experiencing place.

You will notice that the last essay is from someone who was not involved with Homeward Bound at all. She is a young female scientist who wants to go to Antarctica someday. She is the future and I hope this book inspires people like her.

ITINERARY

A JOURNEY TO THE ICE

December 2-3

Drake Passage
Latitude: 56° 43′ S, Longitude: 64° 55′ W
Wind: 13 Knots, N
Temperature: 7° C (8 AM)

We launched from Ushuaia, Argentina and headed 760 miles south through the Drake Passage, a trip lasting 48 hours. Captain Waldemar Wichmann and his crew led us through the Beagle Channel, famously named for Charles Darwin's journey around South America in the 1830's. This area between the 48th and 61st parallels is known as the Antarctic Convergence where warmer, tropical saline currents moving south converge with cold, less saline currents moving north. I took some medication for motion sickness, expecting the worst. But luck was with us - the waves were only 3-9 feet high, and many of us were even able to do yoga without launching off the stern. Our first stop was in the South Shetland Islands.

December 4

Aitcho Islands, Barrientos Island,
South Shetlands Islands
Latitude: 62° 27′ S, Longitude: 59° 34′ W
Wind: 8 Knots, E
Temperature: 0° C (6 PM)

Our first view of Antarctica was the South Shetland Islands, a volcanic arc resulting from oceanic subduction. There are eleven main islands across the southern end of Drake Passage. Some of the major islands include King George, Nelson, Robert, Greenwich, Livingston, Snow, Deception, Smith, and Low.

Although there is a vibrant and fascinating history of the ways in which different countries have tried to claim ownership of the continent (Argentina famously sent a pregnant woman in 1978 to give birth to the first Antarctic "native," for example), ultimately no one "owns" Antarctica under international law. Or, perhaps that is another way of saying we all own it and have a stake in its future.

December 5

Half Moon Island, South Shetlands
Islands
Latitude: 62° 34′ S, Longitude: 59° 51′ W
Wind: 13 Knots, WNW
Temperature: 0° C (8 AM)

We went ashore to Half Moon Island between Greenwich and Livingston. Chinstrap penguins gave us a tour of a 100-year-old abandoned whaling boat while seals lounged around. We listened to the meltwater streaming down the glaciers carrying with it silt, rock, and other debris. Our first landing was calm and contemplative. We bathed in the 24-hour day sun.

December 6

Carlini/Dallmann Station, King George
Island, South Shetlands Islands
Latitude: 62° 13′ S, Longitude: 58° 40′ W
Wind: 24 Knots, W
Temperature: 1° C (8 AM)

The Chilean/Argentinian Carlini Station on King George Island was the first research station we visited. As of 2006, 40 signatories to the Antarctic Treaty established research stations that operate in the austral summer as well as year-round. We toured their seaweed ocean acidification experiments and observed their atmospheric measurements using LiDAR and weather balloons. Changes in the dissolved oxygen content in the sediment and water are linked to stratification and anoxia changes that are used to measure the speed of glacial retreat. Researchers at the station showed us how they studied oxygen uptake and recirculation in the soil in the bay.

The Shetlands are the warmest and wettest region in Antarctica and are noted for biodiversity. Lichen and moss, Antarctica's only vegetation, can be found here. They are considered 'climate change winners' because the rising temperatures and increasing land availability due to glacier loss makes a nice habitat for them.

December 7

Paulet Island, Weddell Sea
Latitude: 63° 34′ S,
Longitude : 55° 47′W
Wind: 2 Knots, W
Temperature: 1° C (8 AM)

We were initially scheduled to land on Paulet Island. The wind made it impossible to anchor, so we floated for a while, watching, waiting. The Northern Weddell Sea forms two gyres in the Southern Ocean that forms a circular, clockwise current that spits ice across the Antarctic Sound. The early part of the season can be especially active. It makes for a dangerous but gorgeous cruise through a scene of tabular icebergs. We made it with a bit of luck. It was a glimpse of Antarctic danger after the gorgeous beaches of the Shetland Islands. We really felt that we were in Antarctica.

The island was covered with hundreds of Adelie penguins and smelled like a dirty aquarium. The island has a history of both danger and safe haven - it is a famous point of rescue for the Swedish Nordenskjold Expedition in 1901 during the "Heroic Age of Antarctic Exploration" when humans first reached the geographical and magnetic South Pole. The Swedish Expedition's aim was to leave a wintering-over party as far south as possible but their ship was crushed by ice. Instead, they were washed up on the island and survived on a diet of penguins and birds' eggs. The ruins of the hut can be seen today.

December 8

Brown Bluff, Antarctic Sound,
Continental Landing
Latitude: 63° 30′ S,
Longitude: 56° 52′ W
Wind: 8 Knots, S
Temperature: -1° C (8 AM)

After our grand plan of reaching Devil's Island was foiled by ice conditions, we instead took our first landing on the continent at Brown Bluff. The bluff is famously rust-colored, and the two miles of beaches provide opportunities to view nesting Adelie and Gentoo penguins. I marveled at how closely the bluffs resembled the climbing meccas of Utah in the United States, except with penguins. As luck would have it, it was sunny enough to lounge on the beach - I even took off my socks and shoes for a suntan. We watched gentoo penguins fanning their wings to cool down by the shore, poising for the cameras along the way.

December 9

Baily Head, Deception Island,
South Shetlands Islands
Latitude: 62° 59' S,
Longitude: 60° 29' W
Wind: 2 Knots, NE, Temp: 0° C (8 AM)
Whaler's Bay, Deception Island
Latitude: 62° 57' S,
Longitude: 60° 33' W
Wind: 5 Knots, SE
Temperature: 0° C (8 PM)

In the morning, we landed on Baily Head where 100,000 pairs of chinstraps constitute an internationally renowned Important Bird Area (IBA). It was a National Geographic moment when two brown skuas attacked and predated a chinstrap penguin right in front of us. But it wasn't television, it was real. Our expedition leader mused that death and the fragility of life are front and center at Baily Head. We also witnessed cape petrels, snowy sheathbills, and Weddell, crabeater, and leopard seals. Snow algae peeked out from under the snow on the black ash ground.

In the afternoon, we made a second landing on Deception Island, the largest of three active volcanoes in the South Shetlands. Volcanic eruptions destroyed the research stations in the 1960s. The island is now abandoned. The area is nine miles in diameter and encloses the Port Foster harbor. Tourists can dig a hole in the volcano-heated soil and wait for the cold ocean tide to come in, creating a makeshift hot tub to warm up in after a polar plunge. The bay is called Whaler's Bay, a historically significant whaling region during the early 1900s. Abandoned oil tanks and blubber rendering machinery sit askew in the disturbed sand, creating an apocalyptic Dr. Seuss landscape. Legend has it that the last volcano eruption caused the ocean tide to "sluice in and out like tea in a saucer," unveiling a graveyard of whale bones that line the bottom of the bay.

December 10

Portal Point, Graham Land,
Continental Landing
Latitude: 64° 30′ S,
Longitude: 61° 39′ W
Wind: 5 Knots, NW
Temperature: 0° C (8 AM)

We cruised through the tight fjords to land on Portal Point in Charlotte Bay. The northwestern part of the Reclus Peninsula, this piece of the Antarctic continent (Graham Land) is only a little toehold of land. These are classic young mountains that make up most of the Antarctic Peninsula. This was an important landing because our expedition leader Greg saw that everyone on board was starting to tune into the place. With each landing, we saw different guises of Antarctica from day to day, and we were starting to "feel" Antarctica. Before that, we were too much in awe of the place.

The weather that day was cloudy. There was not much wildlife, especially in comparison to the penguins of Bailey Head. We hiked up the hill to get a good look around. We allowed ourselves to let loose a little, have a snowball fight, and absorb the peace that Antarctica had to offer.

December 11

Southern entrance of Paradise Bay,
Zodiac Cruise
Latitude: 64° 48' S,
Longitude: 63° 05' W
Wind: 10 Knots, S
Temperature: 4° C (5 PM)

Things changed dramatically. There was an unusual amount of ice for that time of year. Our ship, the 'Ushuaia,' was not a strong vessel and had limited ability to get to where we wanted to go. We tried to test the ice in the southern Gerlache Strait. But we couldn't get through so we changed our plans and tried to go into Paradise Bay. Ice was packed up against the shore and we couldn't get our zodiacs to the shore. Every few hours, our Captain and expedition leaders talked to other ships over radio or email. Although we were able to stream live weather conditions, the resolution of the maps made it difficult to interpret what was happening on the ground. Better information came from other ships. Our expedition leader Greg Mortimer saw the ship he used to own cruise by. He knew they had been further south and, as a stronger ship, were able to get through the entrance of Lemaire Channel. But our smaller ship couldn't make it through. The Captain was nervous. This was a critical phase because we scheduled an appointment at the United States research station, Palmer Station in advance. Because they can't accept many ships per year, they couldn't reschedule.

We were afraid we might miss our chance.

December 12

Southern entrance of Paradise Bay,
Zodiac Cruise
Latitude: 64° 48′ S,
Longitude: 63° 05′ W
Wind: 10 Knots, S
Temperature: 4° C (5 PM)

There were more than nine other ships in the channel. Greg pointed us towards some secret treasures in the Gerlache bays where the large ships don't often go.

There was a low tide and we were able to land on the rocky shore of Danco Island. You could see the tops of penguin heads as they marched in the deep snow. It was their nesting time.

We heard from other vessels that there was sea ice at Wilhelmina Bay. It is called fast ice because it is fastened to land, but it is quite dangerous because it may look solid but disintegrates from beneath. The Captain was against it. Our expedition leader begged the Captain to take a look. He knew that the sea ice was from last winter. It sat between the island and bay in a place that received less sun and was protected from the erosion of currents. The Captain allowed Greg to take his chief mate to go check it out. When they radioed back confirming the ice was stable, the Captain consented. Getting to walk on the fast ice was one of the highlights of the trip.

December 13

Orne Harbour, Continental Landing
Latitude: 64° 35′ S,
Longitude: 62° 39′ W
Wind: 24 Knots, SW
Temperature: 1° C (4 PM)

We traveled up and down the Neumayer Channel trying to get south to make our appointment at Palmer Station. But there was ice. We watched as a bigger vessel passed us and pushed through, but the ice was under so much pressure, it closed the clearing before we could make it. It was a frustrating morning.

We were able to get to Orne Harbor. This was the first time we felt cold, Antarctica cold. It was windy, like Danco Island, but this place felt different. Sarah Charnaud later told me she wrote this in her diary: "the clouds are dark and brooding, hung low over the inky sea speckled with floats of all sizes, portentous of a changing wind." The geography changes appearances from different distances. We had been near here, but it felt like we were in a different part of the world. We didn't let the weather bring us down. We tested out the laws of physics and slid down the mountainside. Who says scientists can't have fun?

December 14

Curtiss Bay, Orleans Strait,
Zodiac Cruise
Latitude: 64° 02′ S,
Longitude: 60° 46′ W
Wind: 2 Knots NE
Temperature: 0° C (2 PM)

We didn't make it to Palmer Station on time. We traveled north overnight and considered going out into open ocean to try to go around the other side. We cruised into southern Bransfield Strait to Curtis Bay in the Orleans Strait to get away other vessels. Curtis Bay is another one of Greg's less frequented treasures. On the landing, we heard explosions of ice collapsing. This is an area where glaciers are active.

December 15

Neko Harbour, Andvord Bay, Continental
Landing
Latitude: 64° 50′ S,
Longitude: 62° 32′ W
Wind: 2 Knots NE
Temperature: 1° C (2 PM)

We went ashore to Neko Harbor in Andvord Bay, a small bay along the west coast of Graham Land. It sits in a fjord among a large amount of ice. An active glacier sits above the bay. But today was calm. We enjoyed the peace of the wildness and hiked up the side. I made a successful polar plunge, this time going all the way in. The water was so cold. It iced as we walked out, cutting some of us on our legs.

December 16

Damoy Point, Dorian Bay,
Wiencke Island
Latitude: 64° 52' S,
Longitude: 63° 48' W
Wind: 30 Knots, NE
Temperature: 1° C (7 PM)

We heard back from Palmer Station!
They said they would be able to change
our date. We headed down Bismarck
Strait in the morning, lining up for
Palmer. The wind changed in our
favor and the ice opened up. We sailed
through the Northern Lemaire Channel
in the afternoon. It just so happened that
the ocean advocate Lewis Pugh was in
a ship near us. He zodiaced over to our
ship and told us about his adventures
swimming in the open ocean to protect
marine areas. You never know what will
happen in Antarctica. You just have to
take life as it comes.

Yuan Goa (right), Associate Professor of Atmospheric Chemistry,
Rutgers University with the author (left)

December 17

Palmer Station, Anvers Island
Latitude: 64° 46',
Longitude: 64° 03'W
Wind: 5 Knots, E
Temperature: 6° C (9 AM)

As if a reward for our persistence, the ship finally landed at Palmer Station, located at Arthur Harbor in Anvers Island. The Station is under U.S. ownership and focuses on marine biology.

Later that afternoon, we sailed to the neck of the Lemaire Channel. But the wind was up to 30 knots there and it was really cold, so we passed it up.

The Scottish-American naturalist, environmental philosopher, and glaciologist John Muir pointed out nearly 200 years ago: "Thousands of tired, nerve-shaken, over-civilized people are beginning to find out that going to the mountains is going home; that wildness is a necessity..." Listening to my recordings to prepare for this book, I am surprised that so many of us used the word "home" to describe Antarctica. This foreign place that was once as unattainable as the moon quickly became a place of familiarity and sacredness.

December 18

Port Lockroy, Jougla Point
Latitude: 64° 49′ S,
Longitude: 63° 30′ W
Wind: 13 Knots, NE
Temperature: 1° C (9 AM)

The station is currently operated by the Antarctic Heritage Trust. It was really a stroke of genius on the part of the United Kingdom to place a post office here. Politically, it acts as an anchor to legitimize Britain's presence on the continent. Economically, the small gift shop funds the UK's Antarctic Heritage Trust. I stocked up on Christmas gifts and sent 100 postcards to people who supported my trip (the postcards arrived in the United States three months later - in March).

We got one last chance at a polar plunge with sunshine and some snow. It was hard to get everyone back on the ship. We had to say goodbye to Antarctica.

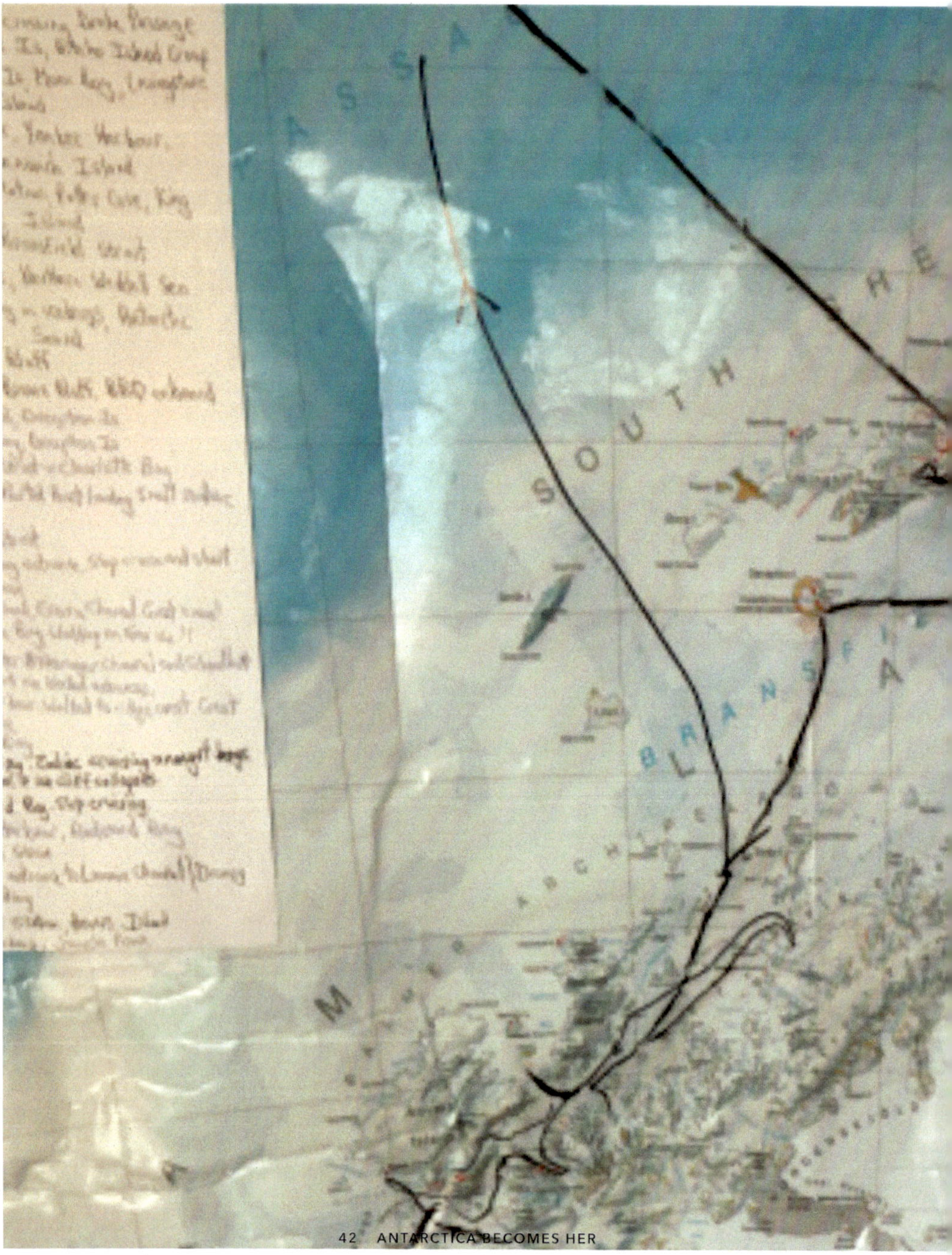
SOUTH

December 19

Drake Passage, Moat Point,
Beagle Channel
Latitude: 55° 00' S,
Longitude: 66° 46' W
Wind: 10 Knots, NNE
Temperature: 8° C (1 PM)

Cruising through the Gerlache Strait again, we watched humpback whales feed on krill around our ship. Albatross and petrels flew around. Antarctica slowly slipped away. Our first morning back in open ocean was bumpy at first, but the waves remained 3-6 feet high.

Greg says the reason why he keeps taking people to Antarctica is because he loves to share the experience. Antarctica becomes our long-forgotten home, one we want to protect immediately after we have been there. I felt the same, like Danielle expressed.

Yes, everyone should see Antarctica! And no, please for the sake of the planet, do not go to Antarctica!

December 20

Ushuaia, Argentina
Latitude: 54° 53′ S,
Longitude: 67° 42′ W
Wind: no data

We made it back to land just in time to see our first sunset in three weeks. Because we were so far south during austral summer, the sun would hover around the horizon. So the sunset was a visual closure to our journey. Now the real work would begin - now it was time to practice our leadership skills and show the world what women scientists could do!

Our expedition leader Greg says that it is important to cruise past Cape Horn on the southern tip of South America. It reconnects us to the land after being away at sea. Antarctica slips away. We become smug, after all, we've just returned from Antarctica, lucky us! The shift from Antarctica to disembarking happens quite fast and dramatically. We are packing, organizing, getting our passports back; there are many things to do. We started thinking about our lives again, our family and our to-do lists. How many emails will be waiting for me in my inbox?

PORTRAITS OF WOMEN SCIENTISTS IN ANTARCTICA

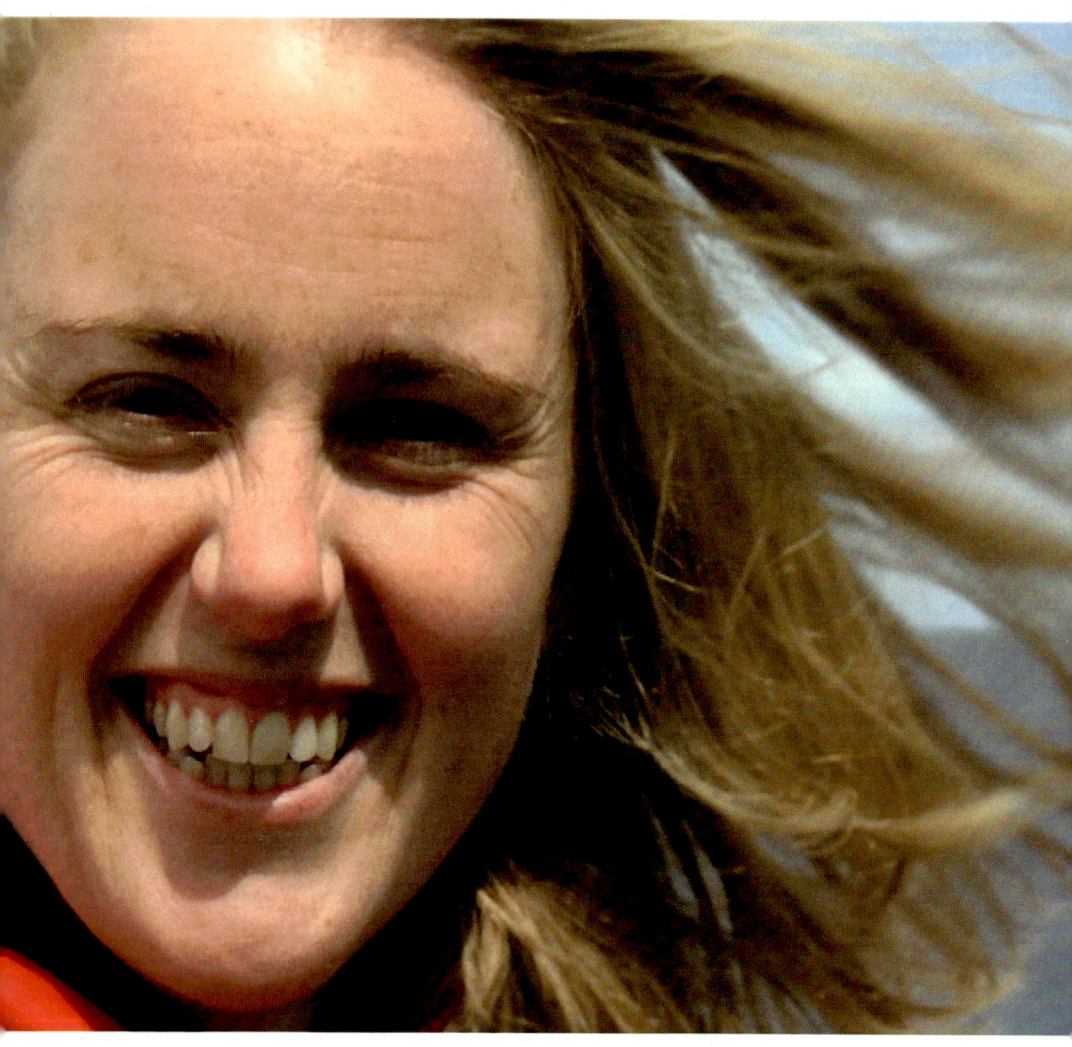

Belinda Fairbrother Tighe
Community Conservation Manager at Taronga Zoo
Sydney, Australia

Kate MacMaster
Programs Director, Peter Cullen Water & Environment Trust Board
Member, WaterAID Australia

Lauren DuBois
Director of Wildlife Rehabilitation-Project Wildlife,
a program of the San Diego Humane Society
Callifornia, United States

Briony Ankor
Leader, Plan B
South Australia, Australia

Marianne Harvey
Director and Principal Consultant, MEGMS
New South Wales, Australia

Sea Rotmann, Ph.D.
CEO, SEA – Sustainable Energy Advice Ltd.
Wellington & Wairarapa, New Zealand

Margaret Barbour, Ph.D.
Associate Professor, The University of Sydney
New South Wales, Australia

Ghislaine Platell, Ph.D.
The University of Western Australia
Western Australia, Australia

Raeanne Miller, Ph.D.
Knowledge Exchange Fellow and Researcher, Scottish Association for Marine Science
Scotland, United Kingdom

Sandra Kerbler, Ph.D.
Division of Metabolic Networks, Max Planck Institute of Molecular Plant Physiology
University of Western Australia
Western Australia, Australia

Colleen Filippa
Teacher / Director, Earth Ed / Fifteen Trees
Victoria, Australia

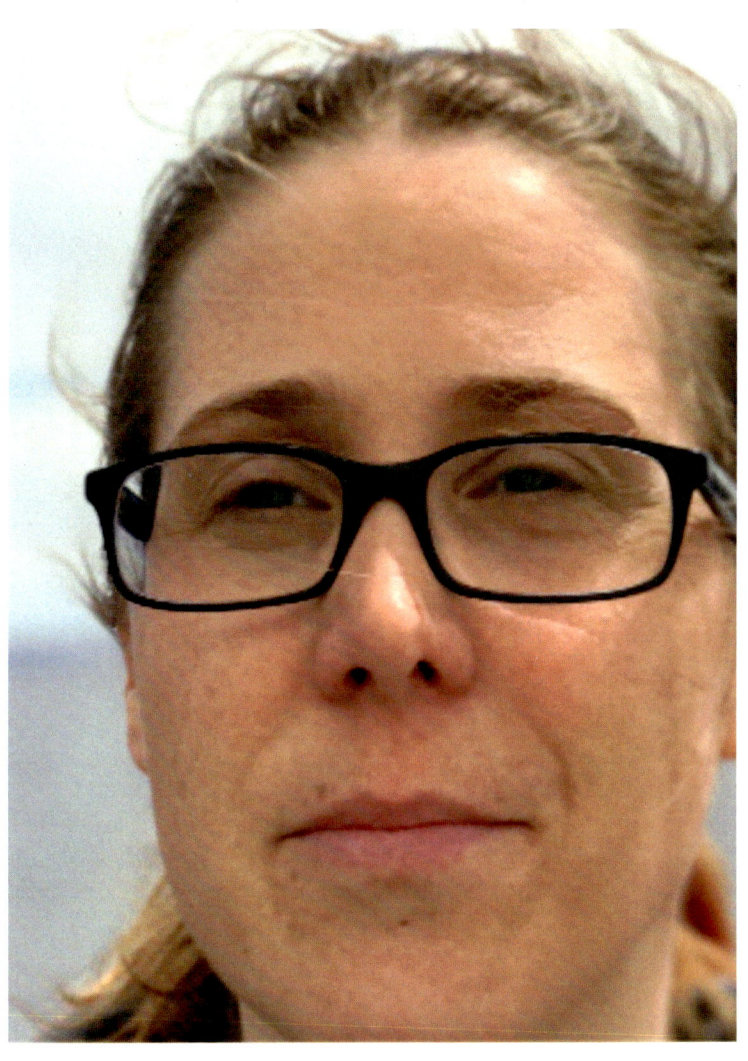

Karen Hawke, Ph.D.
Medical Research Scientist,
South Australian Health and Medical Research Institute
South Australia, Australia

An interview with Danielle Medek

Danielle Medek, Ph.D., MD
Doctor at North Shore Hospital
Aukland, New Zealand

Danielle researched in Antarctica twice before the Homeward Bound trip, once on Macquarie Island and briefly at Mawson Station. Originally from Australia, she earned her Ph.D. in 2008 in plant ecophysiology from the Australian National University. She then received an MD and practices at North Shore Hospital in Auckland, New Zealand.

I asked her about the time she studied plant anatomy and temperature tolerances of grasses on Macquarie Island in 2005 as a Ph.D. student with her supervisors Marilyn Ball and Marcus Schortemeyer.

MH: Thinking back to the time when you were preparing to leave on this trip, what was the number one reason that attracted you to the research?

DM: The interesting research question - These plants live in cold yet moist and stable environments, so we suspected they would be sensitive to temperature perturbations. I had absolutely no idea what it would be like, so it was hard to conceptualize. It was a complete leap into the unknown. But I knew there would be lots of penguins. But it was different. It was very community- and team-oriented. It didn't feel crushingly remote. I was surprised to learn how much our actions influenced each other - Both for safety considerations, but also because the community was so small. Scientists are very self-centered, and are used to burying themselves in their research 24/7. But that doesn't work in a place where everyone relies on each other to eat, drink fresh water, and stay warm.

I was motivated by the remoteness, and the freedom, being on the wide ocean. There were multiple surprising experiences - Moonrise over the Nuggets from the Ham Shack with seals below and the Aurora above. Leaping over rock stacks with an animated geologist while he taught us how they got there. Lying down on the beach, and having curious gentoos surround me. Walking alone across the island. Futile attempts to stick thermocouples under wet leaves, when the graupel hail started to fall on me, rolling around in the sand laughing at the hopeless situation. Carrying a heavy piece of equipment back over the island, and falling face-first into a deep mud puddle, and just surrendering and swimming breaststroke through it.

MH: How did the first trip influence you to go back again?

DM: How could it not! On the first trip, we resupplied Mawson Station and Casey Station. We spent 40 days on the ship, including a week gently drifting through pack ice. We had artists in residence including musicians aboard. We spent hours on deck watching the bergs and wildlife pass. It opened up my horizons, as I met and mingled with so many fascinating people from carpenters to meteorologists, to chopper pilots, to seismic geologists, to filmmakers. The time on the ship allowed me to pause and meditate on my research and my life. On the ocean, with the destination so far away, the journey took on a meaning of its own. The first trip was a scoping mission, and the second was an 11-week field campaign on Macquarie Island in 2006 with my supervisor Marcus Schortemeyer and supertech Jack Egerton. We took plants from the hillsides and grew them next to plants from the shore, and compared their growth and photosynthetic performance under experimental warming. We also compared the photosynthetic performance and leaf anatomy of plants from the shore to the peaks to see how the cooler wetter environment at higher altitude influenced plant form and function.

MH: Can you tell about a time when you were in Antarctica when you felt fulfilled?

DM: Helping the resupply - the refueling and the ship to shore operations. - I felt useful - Not a common feeling for a junior scientist. I'd spent my childhood on boats so was comfortable with that sort of thing. Most proud of? Marcus and Jack were huddled inside the hut at Sandy Bay on our first day out in the field because the wind and rain sounded so dreadful outside, but I was happily swanning around doing photosynthetic fluorescence measurements - It wasn't that wet or windy once you got out there. The weather always seems worse than it really is from the warmth of a hut.

MH: What sort of obstacles or barriers have you faced being a scientist?

DM: I felt like my research community was insular, and an echo chamber. The plants I was studying were being eaten to oblivion by feral rabbits. The cushion plants on the island's peaks were near extinction due to climate change-related weather events. That's when the island seemed remote. It just didn't matter to the rest of the world. There was so much pessimism surrounding science careers. I was daunted by the lack of job security. My second obstacle was when I broke my leg, and complications meant I was on antibiotics for 5 years. I couldn't return to Antarctica as I wouldn't have passed a medical. I had to find a different dream.

MH: If you were talking to a reader of this book, what is the one thing you want them to understand about Antarctica?

DM: Don't go unless you really have a reason. If not, leave Antarctica to itself. Antarctica does not need you.

MH: Do you think you will return to Antarctica?

DM: It's unlikely. I have been very fortunate to go twice for research. The third time felt like I was burning fossil fuels and polluting a pristine environment without giving anything back.

MH: What is the direction of your future research? What kind of research do you think still needs to be done, even if it is not your own? There's still alot of uncertainty about how the Southern Ocean is responding to climate change, and it's such a huge portion of our planet's surface. Understanding the Southern Ocean's dynamics better will help us know what we're in for globally.

MH: Thinking about what you know about Antarctica, how would you empower the reader of this book to lessen their individual impact on the environment?

DM: Unplug and go outside. Once you realize that you're connected to your environment, you'll be far closer to lessening your impact. Stay strong, and healthy. Work on your own wellbeing. When you're stressed and sick, you buckle down to your own personal interest. You don't have time for alternative transport, sustainable food, you just get by with the easiest path to the next day. When you're well inside, you open up to the world, and in turn, the world sustains you.

Britta Denise Hardesty, Ph.D.
Principal Research Scientist
Commonwealth Scientific and Industrial Research Organisation (CSIRO)
Canberra, Australia

Lucy Forde, Ph.D.
Senior Research Scientist, Callaghan Innovation
(The Measurement Standards Laboratory of New Zealand)
Lower Hutt, New Zealand

Andrea Fidgett
Director of Nutritional Services, San Diego Zoo Global
California, United States (originally from the United Kingdom)

Amanda Sinclair
Ph.D. Candidate, RMIT University
Tasmania, Australia

Samantha Grover, Ph.D.
Lecturer, Environmental Science
Applied Chemistry and Environmental Science, RMIT University
Victoria, Australia

Ashton Gainsford, Ph.D.
James Cook University
Queensland, Australia

Johanna Speirs, Ph.D.
Atmospheric Scientist, Snowy Hydro Ltd.
University of Queensland
New South Wales, Australia

Niina Kautto
Research Associate, RMIT University
Victoria, Australia (originally from Finland)

Charlie Hindle
Regional Operations Officer, Environment Protection Authority
Australian Capital Territory, Australia

Mónica Araya, Ph.D.

Mónica Araya told me about the time she ran for Vice President of Costa Rica. Her research on refineries became a regional political issue and catapulted her into the public sphere. She has since founded the Latin American NGOs Nivela and Costa Rica Limpia. Working in the policy sector, she enjoys bringing up the next generation. Despite the fact that millennials might be portrayed as aimless or entitled, Mónica sees them as sources of creativity: "the new generation is very empowered and they will empower themselves," she explains. "Once you help them a bit, then they can do the rest for themselves."

Mónica Araya, Ph.D.
Founder & Director, Fundadora y Directora
Costaricalimpia.org
Heredia, Costa Rica

Ruth Luscombe, Ph.D.
Operations Research Analyst, Queensland Health
Queensland, Australia

Molly Christensen
Hatchery Manager, IMAS
Tasmania, Australia

Elanor Huntington, Ph.D.
Dean, College of Engineering and Computer Science
Australian National University
Australian Capital Territory, Australia

Kerry O'Brienn
Consultant, Regional business development
Queensland, Australia

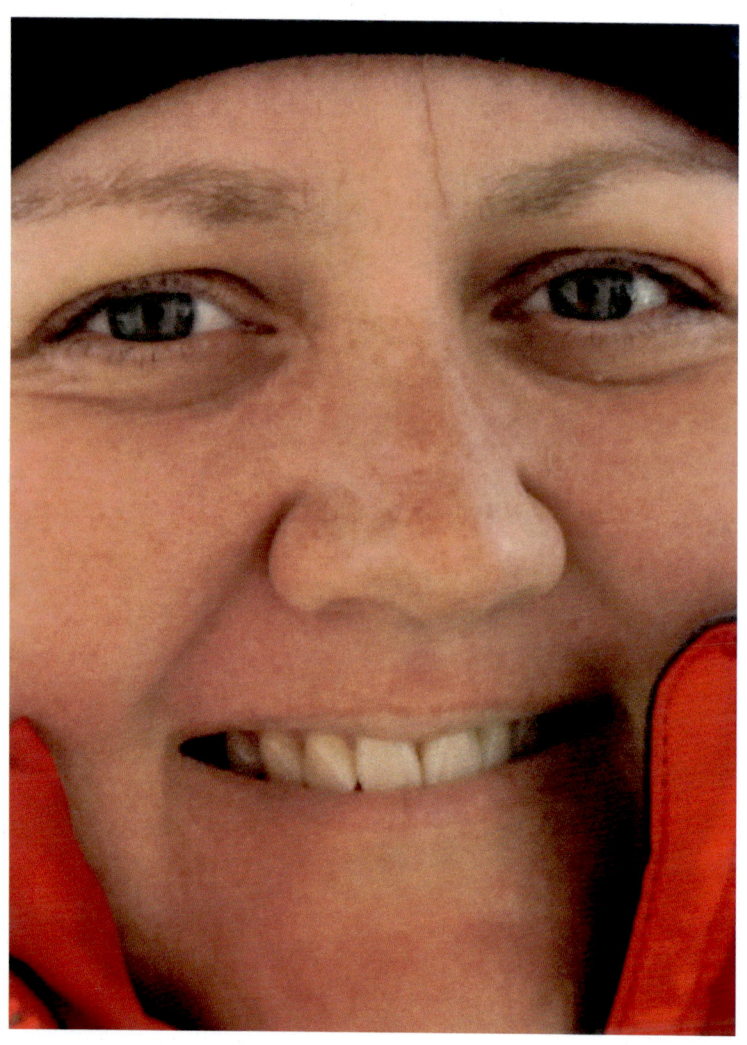

Jennifer Woodgate
Senior Engineer Strategic Planning, Aurizon Limited
Queensland, Australia

Merryn McKinnon, Ph.D.
Lecturer, Centre for the Public Awareness of Science, Australian National University
Australian Capital Territory, Australia

Samantha Hall, Ph.D.
Director and Founder, Rate My Space
Western Australia, Australia

Nina McLean, Ph.D.
Endeavour Fellowship at the Netherlands Institute of Ecology (NIOO-KNAW)
Australian Capital Territory, Australia

Elvira Poloczanska, Ph.D.

Senior Scientist
Alfred Wegener Institute
Bremerhaven, Germany (originally from Australia)

Dr. Elvira Poloczanska is the Science Advisor to the Co-chairs Working Group II of the Intergovernmental Panel on Climate Change (IPCC). It's hard to imagine a woman of Elvira's caliber just starting out in the world as a young scientist. I wanted to know how she got to be where she is now. She told me of one of her first job interviews at a biomedical company. But instead of inviting her because of her already impressive resume, they told her point blank that they just "wanted to see what she looked like" because she was the only woman who applied. She did not get the job. I asked Elvira how she was able to persist despite blatant sexism. She told me she gets her grit from having good mentors. While Elvira is now the mentor for many women, she continues to learn from others. For example, she told me about how she went to a recent meeting and a young scholar pointed out the gender bias in the room. Now Elvira models that colleague at the start of each meeting just to make sure that the issue is always on the table. One thing I learned from talking to the women on this trip is that mentorship played a large role in all of the women's roads to success - being a mentor while taking mentor cues from those younger and older.

Wynet Smith, Ph.D.
Geographer-At-Large and Environment Mainstreaming Advisor, UN Environment
Ontario, Canada

Kathleen Patrick
Science Communications Consultant
Western Australia, Australia

Melinda "Lindy" Fitzgerald, Ph.D.
Professor, Curtin University
Perron Institute for Neurological and Translational Science
Western Australia, Australia

Lauren Sandon
Environmental Consultant, Golder Associates
Victoria, Australia

Amanda Davies, Ph.D.
Dean of Graduate Research School
Associate Professor of Geography, Curtin University
Western Australia, Australia

Renate Egan, Ph.D.
Professor / Leader / Chair, UNSW / UNSW Node of the Australian Centre for
Advanced Photovoltaics / Australian PV Institute
New South Wales, Australia

Lindsay Stringer, Ph.D.
Professor, School of Earth, Environment, and Development
University of Leeds
Leeds, United Kingdom

Cristina Venables
Environmental Water Planner, NSW Office of Environment and Heritage
Australian Capital Territory, Australia (originally from Canada)

Ida Kubiszewski, Ph.D.
Associate Professor, Crawford School of Public Policy
The Australian National University
Australian Capital Territory, Australia

Sharna Jamadar, Ph.D.
Senior Research Fellow, Monash University ARC DECRA Fellow
Monash Institute of Cognitive and Clinical Neurosciences
Monash Biomedical Imaging
ARC Centre of Excellence for Integrative Brain Function
Victoria, Australia

Deborah O'Connell, Ph.D.

Principal Research Scientist, CSIRO
Australian Capital Territory, Australia

Dr. Deborah O'Connell is a Principal Research Scientist of the Commonwealth Scientific and Industrial Research Organisation's Land and Water division. She grew up in Zimbabwe before graduating from school in South Africa.

She says the solution to biases embedded in our culture is really in the 'inclusion' part of 'diversity and inclusion'. She explains, "if a management culture is truly inclusive I believe that this would be reflected in the diversity statistics at senior level, whether it is gender or any other element of diversity."

Leanne Everingham
Water Systems Supervisor, Taronga Zoo
New South Wales, Australia

Alison Davies
Operational Meteorologist, Met Office; Milngave
Scotland, United Kingdom

Glenna McGregor
Veterinary Pathology Senior Resident, University of Saskatchewan
Saskatchewan, Canada

Dyan de Napoli
Penguin Expert / Author/ TED speaker, The Penguin Lady
Massachusetts, United States

Phoebe Barnard, Ph.D.
Affiliate Full Professor, University of Washington
Chief Science and Policy Officer, Conservation Biology Institute
Director, Biodiversity Strategy
Washington, United States
Photo credit: Phoebe Barnard

Amanda Blythe, Ph.D.
Postdoctoral Research Associate, The University of Western Australia
Western Australia, Australia

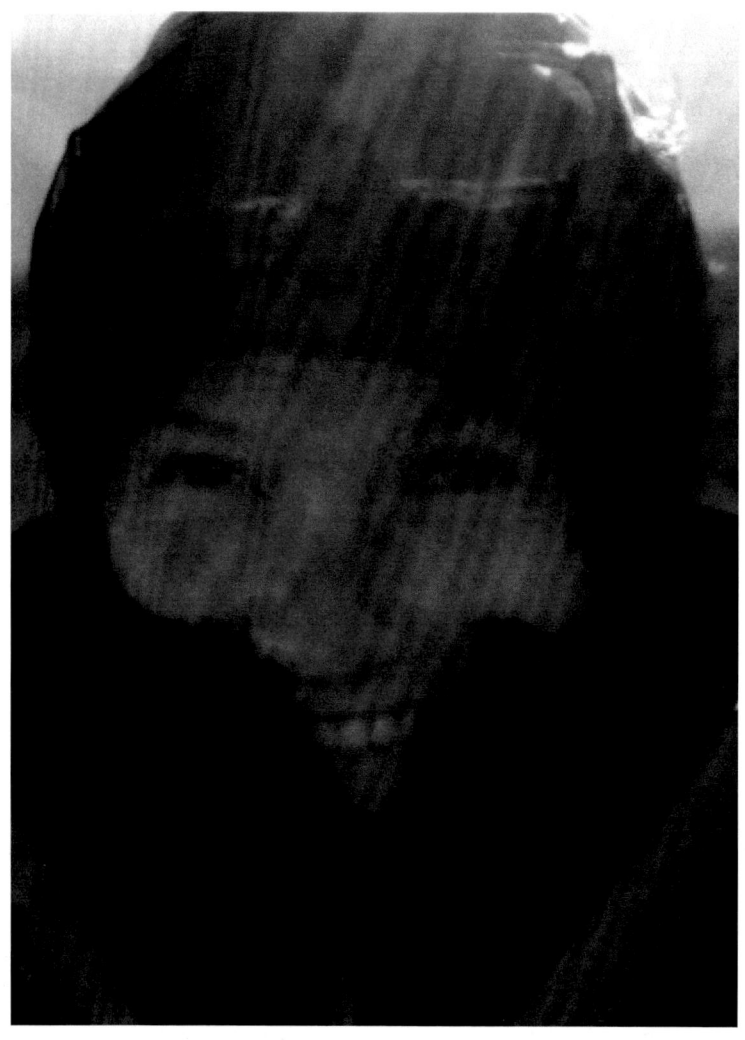

Aimee Bliss
Guide at Tasmanian Walking Company
Tasmania, Australia

Christina Kirsch
Lead Consultant at E2Q / Quattro
New South Wales, Australia

Nicole Webster, Ph.D.
Principal Research Scientist, Australian Institute of Marine Science
Queensland, Australia

Nancy Auerbach

Senior Project Officer, Saving our Species Program, Office of Environment and Heritage
Adjunct, University of Queensland
New South Wales, Australia

"I find it interesting how you talk about these wild places like Antarctica as being sacred, and feeling like our 'home.' That is certainly how I've always felt--I feel so much more like I belong in the alpine, in the Arctic, in the desert, in Antarctica, in the snow, in the sun, in the lightning and thunder, much more than I've ever felt like I belong surrounded by humanity in the concrete corridors of the city, in an apartment block, closed in by the walls and throngs of people around me... I used to do extreme sports--marathon trail running at high altitude, telemark skiing, rock climbing, river rafting--it was in immersing myself in the elements that I felt most alive, and the wilderness was the altar I worshipped. Until I felt like I was taking too much, and not giving back the Earth... which is how I find myself now in threatened species management--trying to help plants and animals and ecosystems and the world I love from going extinct is now something I can't *not* do."

Jessica Reeves, Ph.D.
Lecturer, Geosciences Federation University Australia
Victoria, Australia

Holly North
National Parks Ranger, NSW National Parks and Wildlife Service
New South Wales, Australia

Andrea Robinson, Ph.D.
Associate Dean, Research
Monash University
Victoria, Australia

Joana Picoto Correia, Ph.D.
Research Assistant, University of Manitoba
Manitoba, Canada

Meredith Nash, Ph.D.
Deputy Director of the Institute for the Study of Social Change Senior Lecturer
in Sociology, University of Tasmania
Tasmania, Australia

Sarah Conolly
Senior UG Mine Geologist, Goldfields Ltd.
Western Australia, Australia
(originally from England, United Kingdom)

ESSAYS
ON
SCIENCE

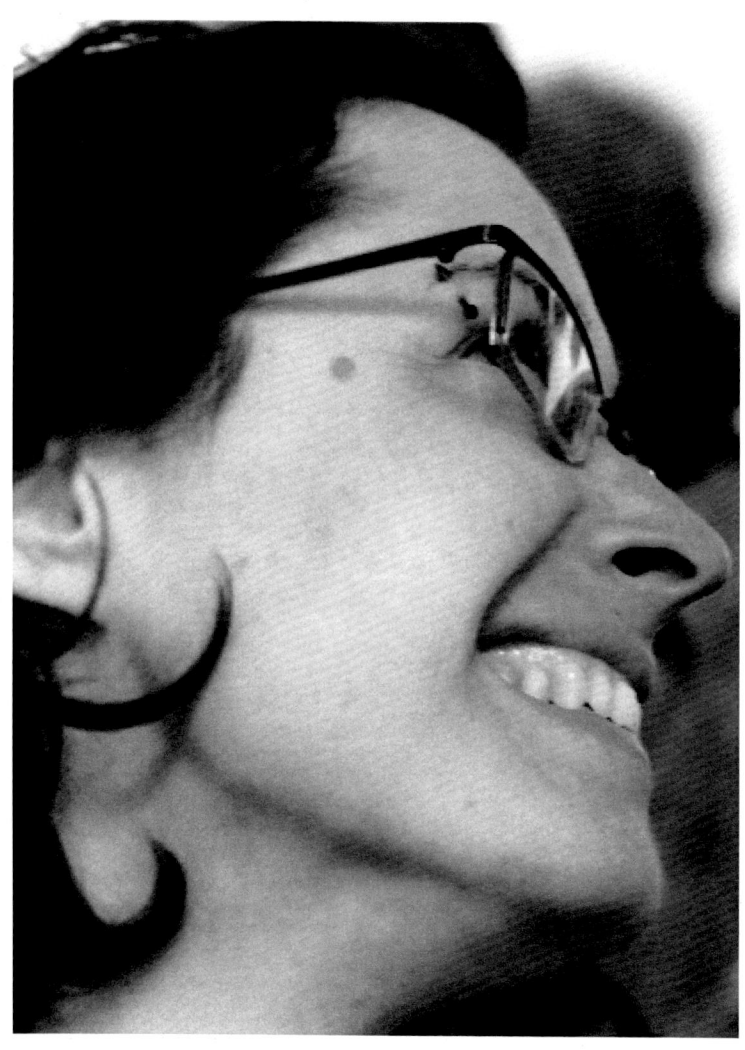

Sarah Brough, Ph.D.
Associate Professor, University of New South Wales
New South Wales, Australia (originally from the United Kingdom)

What do galaxies have in common with icebergs?
by Sarah Brough Ph.D.

We all know that we only see a small fraction of an iceberg and that for each iceberg there is 90% more that we cannot see because it is below the surface of the water. Did you know that the same is true of galaxies?

Before I explain why, let me tell you what galaxies are. Galaxies are enormous cities of stars. Our sun is just one of many billions of stars that make up the Milky Way galaxy. Outside of the Milky Way we see many other galaxies. Astronomers like me have studied millions of those galaxies. When we study galaxies, we start by taking their picture; we take an image of the light coming from those galaxies. The galaxies are so far away that we cannot see their individual stars so the light we see comes from the combined light of the billions of stars. We can use that light to weigh those galaxies and find out the mass of stars present.

In 1933 an astronomer called Fritz Zwicky calculated the mass of galaxies in a cluster of galaxies from how they moved within that cluster. He found 400 times more mass than would be expected from the light coming from those galaxies. He concluded that the extra mass that we cannot see must be dark matter.

Vera Rubin and others in the 1960s and 1970s found further evidence for dark matter by studying how galaxies rotate. She found that the outskirts of galaxies (including our Milky Way) rotate faster than is expected from the light in their stars. This tells us that there must be mass present that we cannot see, otherwise the outer parts of galaxies would fly apart.

We now know that the proportion of this dark matter must be around 85% which means that we only SEE 15% of the mass of galaxies, just like icebergs.

Scientists are still seeking the answer of what dark matter could be. So next time you look up at the Milky Way, have a think about the matter that we can't see.

Shelley Ball, Ph.D.

Founder and Executive Director, Biosphere Environmental Education
Ontario, Canada

Climate change needs visual storytelling: The power of images to change the world by Dr. Shelley Ball (FRCGS)

Towering mountains covered in ice, the incomprehensible variety of blues in glacier ice, the sting of sleet hitting your face during a gale, or the pungent smell emanating from a colony of penguins - this is what it is like to experience Antarctica with all of one's senses. So much of Antarctica is awe-inspiring. And yet watching massive glaciers crumble before your eyes, knowing that climate change is causing this natural process to accelerate, is a sobering reminder of our impact on the planet. I don't think it's possible to experience the frozen continent and not come back a changed person, most especially a person concerned about the future of our planet and all who inhabit it. Despite the fact that nearly 50,000 people per year now visit the frozen continent, Antarctica's remoteness means most people will never set foot there. It's our ability to convey our first-hand experiences in an impactful way that changes people's attitudes and calls them to action. So how then, can we take our life-changing experiences, such as visiting Antarctica, and convey them to others in a way that leads them to care about climate change and our impacts on the earth?

Humans are visual beings. And we are storytellers. For thousands of years, people have used pictures to tell stories, the earliest of them being drawings of animals on cave walls in Indonesia, which are thought to date back to 35,000 years ago. The medium may have changed since then, but today more than ever, a picture really is worth a thousand words. The digital age has propelled us into a world filled with imagery. At times it can be overwhelming, especially with the growth of social media and the abundance of 'selfies'. And yet, few things are more evocative than a photograph with 'wow factor', one that creates connection and elicits an emotional response. It is this power of an image that gives it the capacity to create change. One of the best examples of this is National Geographic. Since September 1888, when their first issue was published, NatGeo has been bringing the world to people. It has educated us, inspired us, and helped create change through awareness. This short video (https://video.nationalgeographic.com/video/00000144-bc15-d540-a5d6-fd157e8b0000) by a collection of incredibly passionate and dedicated National Geographic photographers illustrates the power of an image. "I want them to ask and scream for change," "Photographs can change the course of people's lives," "Never underestimate the power of a still frame," and "Photography can change the world," are just some of the statements shared by this incredible collection of photographers.

With the environmental crises our world is facing, more than ever, we need the power of images to compel people take the actions we need to significantly reduce greenhouse gas production and begin to tackle climate change. Recently the Intergovernmental Panel on Climate Change (IPCC) released their special report (Global Warming of 1.5oC, October 2018) assessing the impacts of climate change with a temperature increase of 1.5o C higher than pre-industrial temperatures. The report, drafted by experts from across the globe, was both sobering and clear. Put the brakes on climate change now or pay the price in loss of human life, livelihoods, loss of species, rising sea levels, droughts, floods, food insecurity, and more. The report was the most dire issued by the IPCC, to date, and concluded that we have 12 years left to put the brakes on CO2 emissions and global temperature rises. The alarm bells have been sounded, but is anyone really listening?

Powerful images of what we stand to lose if climate change is not stopped, as well as evocative images of the impacts of climate change on people's lives already have the potential to make us listen to those alarm bells. When I show photographs of penguins to school kids, it evokes loud and excited 'oohs and ahhs.' My experience is that even the oldest adults cannot be untouched by my photograph of a plump, few day old Gentoo penguin chick peeking out from under its parent's belly and the safety of its nest. Once that excitement has been created by those photos, I then tell my audience how climate change is melting sea ice and that as humans, our daily habits (most especially those of us in wealthy nations) are causing the decline of penguin populations, through the impacts of climate change. Facial expressions quickly change and I can see that these images and the information I just shared has had an impact. Do these photographs and information cause people to change their daily habits? For some, perhaps yes. For most, I believe it at least gets them thinking. I believe that showing these images and sharing stories of our first hand experiences, such as sitting in the midst of a colony of 250,000 penguins, creates awareness. If our messages stick, if they remain in the minds of our audience, then we have the potential to help them change their thinking. And as consumers of resources, we need people to shift their thinking. We need them to understand that even a few small changes in habits and behaviour can have an impact on our world if enough people do it.

So what is it about an image that affects us? Why do some images elicit such deep emotional responses? At the core of it, I think these images evoke empathy. They allow us to put ourselves in the shoes of the person or animal in the photograph. Images can transport us to another time and place. They connect us. More than ever, our world needs empathy. Empathy for our planet and empathy for the living beings that inhabit it. It is through building that empathy that we foster connection and caring and a desire to change things for the better.

In additional to the powerful visual storytelling of National Geographic, new platforms such as Maptia, (https://maptia.com), which give people a place to combine evocative photographs with written stories, are giving a voice to a broader spectrum of people. Also, the evolution of digital video production has also amplified the impact of the 'moving picture.' In just the past several years, documentary films such An Inconvenient Truth, The Last Ocean and more recently, Anote's Ark and The Anthropocene, have brought to us the reality of humans' impact on the planet and provided motivation for action. Perhaps one of the most important documentaries of recent time was Chasing Ice, an account of James Balog's Extreme Ice Survey. Balog and his team set up 43 still cameras on glaciers around the world. These cameras were programmed to take photographs every half hour of daylight, for 365 days. The photos that resulted showed the extent of ice retreat in these glaciers around the world. It was the first time that such unequivocal and convincing visual data existed to show the extent and pace at which glaciers were melting. I think Balog and his team did more than any other initiative to convince the world that the glaciers were melting at an alarming rate and that global climate change is real.

Using social media to amplify the messages from still images and documentaries is critical. Just as the cause of environmental degradation is largely due to the extreme number of Homo sapiens that populate planet earth, the solution also lies in the numbers. The more people we can reach with our messaging of the need for action on climate change and other environmental issues, the more likely the change will happen. Social media has given us a tool to extend our reach and to influence on a global scale.

How can each of us play a role in using visual storytelling for change? Few of us are of the calibre of those incredible National Geographic photographers. And even fewer of us have the skill to produce a documentary film. But each of us, by virtue of being human, is a storyteller. We relate our experiences of life to each other, and by doing this, build connection. Now is the time to capitalize on this truly human trait. And as such visually oriented beings, each one of us can use images to help us tell our stories, whether it is of what's happening in our local community or what is happening in one of the remotest places on the planet, Antarctica. We can use our storytelling tools - smartphones, tablets, digital cameras - to capture images and to combine those with our personal stories and first-hand accounts. That is how we build connection. And it is how we foster change. Find an issue you are passionate about. Go photograph it. You don't have to be a National Geographic photographer to do that. Capture the images in a way that speaks to you. And then share that image with written or spoken words, on any platform you choose. We all have the power to influence and to create positive change, if we choose to. I hope you will choose to.

Betty Trummel

Betty Trummel
Elementary Educator and Northern Illinois University adjunct - retired
owner – The Science Roadshow, education outreach specialist
Illinois, United States

Education and outreach focus: Science communication by Betty Trummel,
The Science Roadshow

Becoming a teacher…first choice; science educator…not so much

I always wanted to be a teacher, but my early days as a student and training to be an educator did not fill me with a love of science. Any science courses I had were basic and rather uninspiring. Despite this, I entered into the teaching profession in 1978…with a huge amount of enthusiasm.

A shift to focus on science in my elementary classroom began three years into my career, when I attended a weeklong, summer environmental conference. I met other educators who led field trips teaching about wildlife, plants, geology, and other aspects of the environment. I thought back to my life growing up on a small farm, and how I spent the vast majority of my time outdoors. I was hooked on science! Most critical, I found the camaraderie of educators and friends with similar science interests. That became a catalyst in both my professional and personal life.

Fast-forward about 12 years…

While continuing to teach elementary school, I worked for several weeks each summer at the very conference that had inspired me. Making personal connections was a highlight and provided me with a form of professional development and mentorship. I looked outside of the box, outside of my school district to enhance my science education skills. Science simply wasn't viewed as an important subject area for professional development, but I was hungry to learn more and implement new projects to incorporate science into my classroom and expose children to hands-on, experiential learning.

I found I had the ability to expose students to careers related to science and change their perception of what a scientist actually does. Possible science careers opened up before our eyes!

Meanwhile, I earned a Master's Degree in Science/Outdoor Education in 1991. And, the stage was set for an exciting new chapter in my life when I was awarded the Presidential Award for Excellence in Elementary Science Teaching in 1996.

An explosion of science communication and outreach begins…

While attending the Presidential Science Award ceremony, a representative from the National Science Foundation (NSF) told me about a unique teaching opportunity. I applied and was chosen for Teachers Experiencing Antarctica and the Arctic (TEA). Soon I was headed to McMurdo Station, Antarctica to work alongside geologists drilling into the sea floor to retrieve sediment cores: The Cape Roberts Project. It took my science learning and enthusiasm to a whole new level…working as part of a multi-national research team, following real-life scientific research in action and sharing it with a broader educational audience through daily journals, emails, and photographs.

TEA was set up to immerse teachers in a research experience as a component of their continuing professional development, and to bring polar research into classrooms in innovative ways. It was an incredibly motivating opportunity for me, and I forever will be grateful to the NSF for selecting me for this experience. I'm also grateful to the huge team of Cape Roberts scientists who welcomed me into their project, taught me all they could, and supported my communication efforts surrounding their science research and the overall project. I know that when there is a true team effort of scientists and educators, it's a win-win situation. Everyone learns, everyone shares enthusiasm for science!

The Cape Roberts/TEA experience was key; even more important it fired me up to share the work of scientists with students, teachers, and general public long after the actual time in Antarctica. Presentations, teacher workshops, the National Science Teachers Association annual conference, and networking with educators around the world became part of the fabric of my everyday life and teaching. Teachers are looking for the professional development that inspires them…I certainly found that!

Beyond textbook science…

In 2005 a new international collaboration of earth scientists planned the ANDRILL (Antarctic Drilling) Project. Following the work of the Cape Roberts Project, ANDRILL's goal was to obtain sediment cores for multi-disciplinary study. I applied for a spot to be one of six educators on the education outreach team. This collaboration of scientists and educators was several steps beyond my first geology research experience, mainly due to the fact that this time I was part of a team of educators from four countries. I wouldn't be "flying solo." We would all be instrumental in sharing ANDRILL's work and developing curriculum materials.

During the three-month ANDRILL project (2016), I was totally immersed in the science. With improved technology, I was able to share our work with an even broader educational audience. This innovative type of program infused learners with excitement. There were numerous opportunities to connect with the process of science and transfer that to classrooms around the world. It had a tremendous impact on me as an educator…knowing that I would step up to continue creating and delivering education outreach materials and presentations for teachers and learners of all ages around the globe!

In addition to immersion in the science teams, there were on-going opportunities to advance our content knowledge in many areas of geoscience, as ANDRILL scientists gave daily lectures. We were welcomed into the ANDRILL scientific community as an active part of the project, and this was an excellent learning experience for educators and scientists alike. Immersion enabled educators to connect and share the scientific work being accomplished. It benefited those around the world who learned about ANDRILL as it took place…being able to ask questions, read daily informational blogs, take part in video and teleconferences, and see photographs of scientists at work in Antarctica.

Another key influence, I made an important connection with one particular educator, Matteo Cattadori, from Rovereto, Italy. We have continued to collaborate on projects, long after the ANDRILL experience. It's inspiring to work with an educator who shares a similar commitment to science. It makes us strive to be better teachers! Just two years ago, Matteo asked me to be a part of an educational journey with 20 of his high school students....18 of whom were girls. We traveled to Arctic Svalbard for an intense 10-day tour, focused on science and learning. All of the students, along with Matteo, have continued to teach others through the many take-aways they created as part of the experience. Articles, a documentary film, presentations, art work created while in Svalbard,

My varied experiences have provided a unique window to the world of science and communication, and shaped my teaching and how I've delivered earth and life science information during presentations around the world. Meaningful professional development improved my scientific content knowledge, provided participation in the process of science research, and gave me the tools to carry forth the excitement of science to students and educators. For an elementary teacher to have an Antarctic geoscience opportunity is incredible. Having the privilege to represent three science teams and the NSF on so many levels is monumental. I've touched thousands of lives, and they have touched me as well. What an honor to be a science educator and communication specialist!

The next chapter...

From a science and communication outreach perspective, being able to share the work of scientists in action is critical. This includes professional development of educators and outreach to a broader educational community as a fundamental goal of all of my outreach efforts since 1998. Positive outcomes from this type of integration have extended far beyond any field season, and have had a lasting impact.

In regard to Homeward Bound, I have found additional mutual support of educational outreach initiatives, an added multi-cultural (and sometimes multi-lingual) perspective, better dissemination techniques, the organization of educational events (with each other and in our own countries), and a significant enhancement in my ability as a science communicator and educator. As we develop international partners in education, it broadens our world perspective and brings learners and classrooms around the world together in a common learning environment. Connecting scientists to classrooms improves and updates our content knowledge, goes beyond textbook science/delivery, and helps demonstrate an interest in lifelong learning as a preferred channel for the transfer of information.

After thirty-five years, I retired from classroom teaching in June, 2015. The next chapter includes my own small business, The Science Roadshow, which is dedicated to promoting lifelong learning in science and technology. Goals: keep teaching, be part of new projects and adventures, and stay involved in education, outreach, and science communication.

I have a responsibility to keep science alive in classrooms, to open new doors to learning. I must continue to find innovative approaches, hone my professional skills, acquire new content knowledge, and transfer the impact of science research to education and society. Professional development doesn't stop...it's a lifelong goal.

An Interview with Nicola Gaston

One might not expect a nanotechnology expert to wax poetic about the lack of inclusivity and intersectionality in academia, but that is what I got when I spoke with Dr. Nicola Gaston, Associate Professor in Physics at the University of Auckland. Her book, Why Science is Sexist, was picked up by the Royal Society of New Zealand. In the same year, for the first time in their 150-year history, they awarded half of their fellowships to women, including the first Maori women. The book has elicited personal stories from women all over the world, leading her to posit that women are not so much pushed out of academia as attracted by more rewarding opportunities elsewhere.

MH: You mentioned earlier that had gotten some emails from women who have left academia and they shared their stories. Is there a common thread from what they've shared?

NG: The most disturbing one that I've picked up on is guilt. Some women feeling that they're letting their side down by leaving. Or that they're letting young women down by leaving. And that's probably the thing that most bothers me from the tone of what I get from people. These are women who are smart and have been successful, but they have had a change of personal circumstances. Or they have had some other great opportunities come up. So often, it's really for good exciting reasons that they're leaving academic science. But they feel guilty and I find that really wrong. It's sort of this extra burden on top of everything else that just shouldn't be there."

Nicola Gaston, Ph.D.

"One of the interesting things," she says, "is that women are much more likely to develop other skills. I think women see themselves as able to do more and see opportunities for themselves outside the system."

Associate Professor, Department of Physics at the University of Auckland,
Co-Director of the MacDiarmid Institute for Advanced Materials and Nanotechnology
Auckland, New Zealand

Tracey Gray

Environmental Science Educator
Moonbird Education
Victoria, Australia

Oceanic connections and agents of change
By Tracey Gray

As children, we gaze into rock-pools with wonder and awe, capturing a glimpse of the life that is found on the edge of a vast and mysterious ocean. We only truly see the edge of the ocean - the massive body of salty water that covers 70% of our planet. All of us have different experiences of the ocean. Maybe we recall the first time we saw the glistening water, how the sand felt, or perhaps the cool connection of the water as it laps the shore. Of course, the ocean can be harsh and powerful. On a gloomy day the ocean appears to have emotions, reflecting characteristics that are wild and untamable.

The ever-changing tide brings new wonders of its own. Ocean life that has lived in the depth of the sea unseen, or something that has drifted and floated from a different continent is an ever present reminder that we are all connected to this deep and powerful life force. The ocean itself is the planetary temperature regulator. It feeds us and it controls the weather. The ocean is a carbon sink and a source for creating vital oxygen. We are intrinsically connected to this life force.

Our role is a challenging one. On one hand, we love our oceans. We feel spell-bound and intrinsically connected to it. But we have not treated our oceans well. Globally, we are watching island homes being swallowed by the warm seas that surround it. We are seeing an increase in plastic in our oceans. We listen to staggering stories about the amount of plastic particles it sea water. Plastics are filling fish, seabirds, and us. Our marine resources are being depleted as fish catches decline.

Our daily impacts are changing our oceanic environments. Is the reason why we let this happen because we only see the oceanic edge? Is it the expansive liquid space that enables this reality to remain invisible? How can we affect change? How can we become more connected to this open, vast, and expansive liquid space that supports us all?

We need to look deeper to make the 'invisible' visible. Science helps us uncover these mysteries of the ocean by studying the dynamics of ocean environments and the marine species that inhabit them. These scientific investigations are vital for our understanding. Interpretation and communication of science to enhance our connection is equally important.

Art science collaborations bring the oceans so we can see for themselves. It provides a narrative of ocean connection that allows us to interpret nature. Story telling of the oceans, its beauty and wonders allows us to explore issues. Personal connections bring home truths.

But we need more. We need community engagement to translate stories to action and change. We need to be the changemakers, and we need to stand as everyday changemakers together - as scientists, as leaders, as decision makers, as adults, as children, as people, as one.

Communities can be powerful if we believe in our impact. Working with passionate people connect to their cause to the sea is engaging. Each everyday decision can make change - from picking up plastics to refusing a plastic straw. Our actions can be making a switch, actively joining on ground teams, listening to our oceanic hearts, listening to our children, and helping to protect the seas. Our attention needs to focus on innovators and inspiration. We need to lead the future of cleaner oceans, cleaner connections, and cleaner collectives.

Clean ocean communities work together to follow the source of the ocean pollution. Communities like this come together to conduct inspirational programs based on philosophical ideas. The founder of Tangaroa Blue, for example, worried that 'If all we ever do is pick up plastics. That's all we will ever do.' We need to do more. We need to inform government, agencies, and the commercial sector that this is not what we want for our oceanic future. We want to see the change and be a part of it.

My advice to you is to choose to be a part of on-the-ground social and environmental movements that affect change. Choose to inform, choose to connect, and chose to engage. Choose a plastic-free future.

When we pick up plastics with schoolchildren, we teach them to ask: How did this plastic get here? What is the story of this item? How did it get away? Was it a thoughtless action? Was throwing it on the ground all too convenient? When we throw something away, where is 'away'?

Ultimately we need to change. We need to change our simple, everyday actions for a healthy oceanic future.

As children, we gaze into rockpools with wonder and awe. What do we see as our oceanic future?

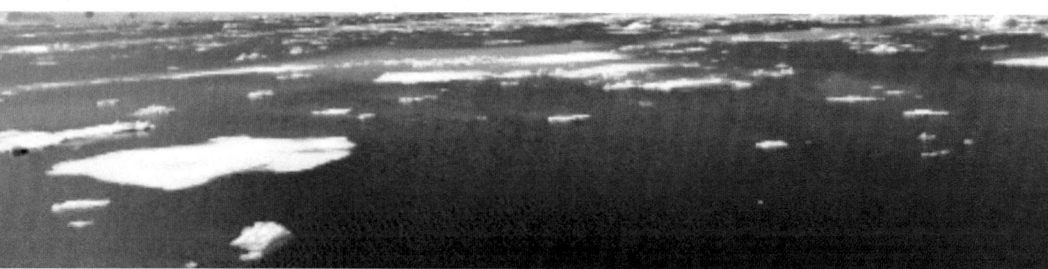

Fern Hames

Science Manager, Communication and Collaboration
Arthur Rylah Institute for Environmental Research
Department of Environment, Land, Water and Planning
Victoria, Australia

"We're on the fast ice, and at a rather precise scale we're a hundred metres, maybe, from the edge where the zodiacs dropped us off… which was an extraordinary thing, to step from water onto this layer of ice. … I am as unknowing of my geography as I've ever been in my life. I've always known exactly where I am, and here; I've just abandoned all of that… I know I'm in Antarctica. I know I'm on the Antarctic Peninsula. I know we were in the Gerlache Strait this morning. But I've no idea what that bay is called and it doesn't matter what it's called…

I was thinking about how almost every place I've heard of in Antarctica is named after a person…Where's 'Albatross Point', and 'Granite Peak'? I suppose they must be here somewhere but it's like we've tried to people a place which doesn't have Antarcticans. …So I love this place because it's so unpeopled, – apart from us, but you can kind of ignore this temporary presence of us…"

On reflection
by Fern Hames

I had dreamed of Antarctica for decades.

I had wanted to go to Antarctica, breathe Antarctica, see, feel and smell Antarctica, for a long time.

When I was accepted onto the Homeward Bound program, my first reaction was disbelief. Then, exhilaration. Finally, Antarctica.

I'd first become a bit obsessed with Antarctica in 1981, doing an Honours thesis on Antarctic and Australian algae and mosses, but studying my supervisor's samples. Unable to get a place to visit there myself, I'd changed careers, and ended up working on freshwater fish, and then participatory conservation, engagement, and science communication. Increasingly, I'd become interested in how we connect with nature, and how we encourage people to act for nature. People make powerful connections with places when they experience them directly, especially in intentional ways; when they are actively noticing or exploring things around them, and truly connecting with the place. I'd observed this in many places, including in the deserts of central Australia, with teams of citizen scientists, and seen people transformed by deep connection with those remote, arid spaces (which I thought of as 'redspace'). I wondered how similar or different it would be, connecting with the wild, white, desert landscape of Antarctica (which I thought of as 'whitespace')?

When I studied the program for our Homeward Bound voyage, I realized it was going to be pretty intense. It looked like there wouldn't be much time for focusing on really 'being' in Antarctica. I wanted to encourage people to make that time. I wanted to encourage them to intentionally pay attention to the extraordinary place we had the privilege to visit. I also thought spending some time being actively mindful in this awesome place would be helpful to people; helpful in dealing with that intense program, being disconnected from home, being amongst 80 strangers for three weeks, potentially being sick, tired, cold, anxious.

As part of the Homeward Bound program, a group of us were exploring transdisciplinarity. Here was an opportunity to integrate attention to place, in a mindful way, within our project work. I created a draft booklet, called 'On Reflection' and our group refined it, and provided a copy to each participant on board. The booklet invited people to write or draw what they could see, hear, feel, smell and taste in any given place, and record how they were feeling. In the Introduction we wrote: "Finally, we are here. We are in a wild place; remote from the multiple absorptions of our lives at home. We are amongst a rich community of extraordinary women; from a range of disciplines and places, with much in common. We want you to make the most of this opportunity; of the place and the people. We will be busy on the ship (and off it too). We believe it's important to remember to breathe (you probably agree). We are giving you this little booklet to help you step back, breathe, reflect, and really notice what is around you. How are you feeling?

What can you see, here actually IN Antarctica? What can you hear? What can you smell, taste, feel? Pay attention to it. Write it down. And now to the people – who have you connected with? Do you think there's an opportunity for working together? Think about it. Write it down. Breathe. Reflect. Be here."

Some participants embraced it, and some ignored it, but the offering was there. For me, I held tight to the idea and worked hard to find those few precious minutes where I could truly BE in Antarctica, and weld it to my soul. My entries quickly shifted from simple lists of what I could see and hear to wider, journal type reflections on being in that space. When I re-read my entries from the day we visited Neko Harbour, on 15 December 2016, it takes me straight back there:

"Feeling simultaneously relaxed and exhilarated. The sea is 'breathing' its tiny 'tsunamis'; like a kind of sighing; tiny little lifts and shifts up and down the tide line; in and out; in and out, and it's almost meditative, and incredibly calming. Feels like we are in a world of our own; surrounded and disconnected by a wall and dome of low cloud. There is a tiny piece of apprehension every time the glacier booms and calves; the cloud limits our view to the distance, so we can't see what's out there; is a bigger tsunami coming? I go for a polar plunge; take off my multiple layers, surrounded by a silent snowfall of thick fat fluffy flakes, wait for Sarah, then we both run in. My feet are numb within 2 seconds, as they try to gain traction on the pebbly substrate. The water is covered in brash ice and it also seems like the sea itself is forming an ice layer on top; its freezing up – I push through it; it is heavily crusty and crunchy and viscous – a bit like swimming in a not-quite-processed salty slushy. It is SO cold. We squeal. Try to breathe. Gasp. I run out, feet freezing. My legs sting; I look down and have rows of thin horizontal cuts from running through the ice; they are bleeding, and sting from the salt and cold. An intense feeling of exhilaration, of being so fully and completely alive, and present."

The sense of being alive, and present, and in awe were common themes amongst many of the entries by the women in our group. There were also common themes of connectedness and disconnectedness, and a sense of privilege at having the opportunity to be in Antarctica.

Our project group evolved the active reflection practice further into one on one interviews during our landings. This triggered an extended exploration by our team into creating a bridge between natural / physical and social sciences. It further evolved into a writing fellowship at the Rachel Carson Center in Munich, Germany, and a follow up writing retreat at the Arthur Rylah Institute for Environmental Research in Melbourne, Australia. Three publications are in development. This is a powerful example of the core of what Homeward bound is about; bringing together women with a background in science, to collaborate across the world, change the way we do things, and to make an impact -impacts on ourselves, our networks, and -we hope- on conservation practice globally.

As one of my own booklet entries reads: "In awe. Peaceful, grateful, full of intent to do good…"

An interview with Nicole Hellessey

Ph.D. candidate, University of Tasmania (UTAS)
Tasmania, Australia
Photo credit: Carol Devine

MH: Can you tell me about the first time you did research in Antarctica?

NH: I was a first year Ph.D. student. I went to the Ross Sea Region of East Antarctica with CSIRO and UTAS through an international program called GO-SHIP for the P15S voyage on the RV Investigator (an Australian marine research vessel). We were working on the Genomic analysis of the water column

MH: Thinking back to the time when you were preparing to leave on this trip, what was the number one reason that attracted you to the research?

NH: I had been studying Antarctica for 9 years and I had finally been given a chance to go and see this place I knew so much about. I had to see it and experience it for myself.

MH: Thinking back to the time when you were preparing to leave on this trip, what did you think the experience was going to be like?

NH: Exhilarating, tiring, fun, physically demanding, stressful, cold, full of firsts.

MH: What motivated you to go the first time?

NH: I wanted to take the chance to go south as I didn't know if I would ever get the opportunity again. For me, going to Antarctica for research was physical proof I had 'made it' in my research career. I was being taken seriously as a researcher.

MH: Did anything surprising happen?

NH: So many things! Equipment failure, rough storms, creative fixes to problems we didn't expect and "illegally" streaming Season 6 of Game of Thrones every week with the whole crew.

MH: Looking back on the trip now, how did the first trip influence you to go back again?

NH: I knew I wanted to go as soon as I saw my first iceberg, my first pod of humpback whales, the first time I saw ice forming on my eyelashes. I was hooked. This is what I had spent years learning about, teaching others about, researching and know I didn't just get to see it but I could go back and tell people exactly why we should save and protect it. I'd seen the damage we had done with my own eyes and I was more motivated than ever to try to be a voice to help stop it.

MH: Can you tell about a time when you were in Antarctica when you felt fulfilled?

NH: I was sitting on the bridge of the RV Investigator and spotting icebergs with the crew while we talked about discontinued 1990s candy. I was so happy and content having just finished a 12 hour shift of running samples and filtering water and was here talking to people I would have never met if I wasn't there on this ship in the middle of the ocean helping look for icebergs. It was surreal but so fulfilling.

I went again in the second year of my Ph.D. program to the West Antarctic peninsula in December. I'm most proud of my quick thinking of how to solve an issue we had on board the RV Investigator. The water samples we had collected at depth (~5 km down) were 2°C at depth but the air at the surface was -17°C. The water froze instantly on contact with the air, so we couldn't get it out of the Nisken bottles of the CTD (an oceanography instrument). I ran to my cabin, grabbed a hairdryer and ran back to the sampling room. I sat with the hairdryer on low for 45 seconds on each of the 36 Nisken bottles but we got our samples and the Voyage Manager congratulated me on my quick thinking. I was super proud of myself for the rest of the day.

MH: What sort of obstacles or barriers have you faced being a scientist?

NH: The biggest obstacle I've faced was people patronizing me or overlooking me because I'm young and therefore must be inexperienced. Except I'm not. I've been studying science (in particular marine and Antarctic science) for over a decade. But a lot of the older researchers ask their peers for ideas or solutions, but never graduate students despite us being in the lab and field quite often more recently and more frequently.

MH: If you were talking to a reader of this book, what is the one thing you want them to understand about Antarctica?

NH: Antarctica and the Southern Ocean are the lungs of our planet. The algae that blooms every spring, when the sea ice melts, produces huge amounts of oxygen and takes up vast quantities of carbon dioxide. If Antarctica and the sea ice around it melts then we will be starving our planet of oxygen and the rate of carbon dioxide residing in the atmosphere will increase enormously.

MH: Do you think you will return to Antarctica?

NH: Yes, definitely. I want to conduct research on the Antarctic ecosystem until the day I die. So hopefully that means I'll get to go back to see and collect samples from the ecosystem again.

MH: What is the direction of your future research?

NH: I would like to see a study done on the impacts of climate change on salp physiology and population numbers.

MH: If someone asked you how they can go to Antarctica, what advice might you have for them?

NH: Depending on if they studied science or not I would recommend tourism as a sure fire way to get there or if they wanted to do research I'd encourage them to talk to other researchers in their field of interest to see what projects might be coming up that they can get on board with.

MH: Thinking about what you know about Antarctica, how would you empower the reader of this book to lessen their individual impact on the environment?

NH: I would ask them to question what they do in day to day life that they don't need or that they can trade for something more environmentally

NH: I'd love to continue researching the ecosystem in Antarctica particularly the energy and omega 3 pathways.

MH: What kind of research do you think still needs to be done, even if it is not your own?

friendly. E.g. using the bus or walking vs taking a car or eating at home vs take out once a week or having shorter showers or turning the heating down 2°C in winter. Small but real changes over a year and a decade that will make a lasting difference for the Earth.

Krill and the krill fishery
by Nicole Hellessey

Antarctic krill (Euphausia superba) are small crustaceans that grow to roughly 6cm in length and 1 gram in weight. They are the most abundant species of krill in the Southern Ocean and provide food and nutrition to all manner of species; from whales and seals to penguins, albatrosses, petrels, squid and fish. Krill themselves eat a variety of small food items including bacteria, phytoplankton, zooplankton (like copepods), fish larvae, marine snow and they are also known cannibals.

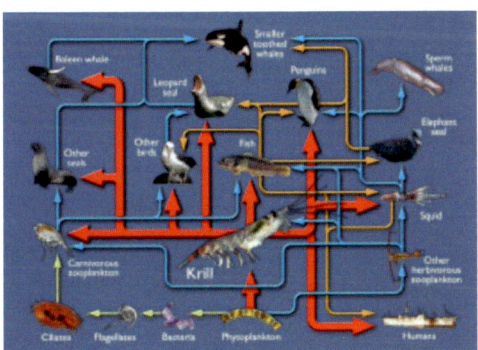

Krill are therefore at the centre of the Antarctic and Southern Ocean ecosystems' food web, with everything larger than a krill eating them and krill themselves eating everything smaller than them. This kind of ecosystem is called a wasp-waisted ecosystem and due, to this pinch point, many people are concerned over the levels of commercial krill harvesting that takes place in the South Atlantic Ocean, or Scotia Sea.

The Antarctic krill fishery started in the mid-1970s and was originally circumpolar in scale. However, after going through a massive growth in the late 1970s before declining due to the collapse of the USSR in the early 1990s, the majority of the krill fishery moved to greener pastures in the South Atlantic where ports were closer and krill populations were larger and more consistent year round. The fishery has consistently harvested krill at low levels ever since and even with the small increase in interest in harvesting krill, particularly for their omega 3's for human consumption, the fishery still harvests krill at a much lower level now, catching roughly 250,000 tonnes a year, almost exclusively from the Scotia Sea. However, this is only

a fraction of the estimated 60 million tonnes of krill in the Scotia Sea alone, and the 400 million tonnes of krill in the entire Southern Ocean. In fact, the krill fishery is often regarded as the only fishery in the world not in decline or collapse due to its low harvest levels and strict management policies.

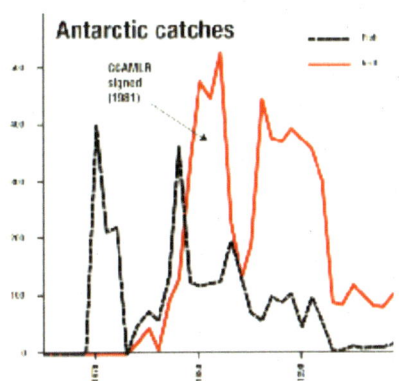

The krill fishery is regulated by the Convention for the Conservation of Antarctic Marine Living Resources or CCAMLR. CCAMLR uses an ecosystem based management approach to govern the levels of krill, fish and squid caught in the Antarctic and Southern Ocean. This approach monitors predator populations for changes as well as using the best available science on the Antarctic ecosystem to run models every year of the predicted krill population and recruitment.

These models are updated every year and new fishing catch limits are set based on the models outputs.

Information on krill health, growth, reproduction, diet and even their resilience to climate change is

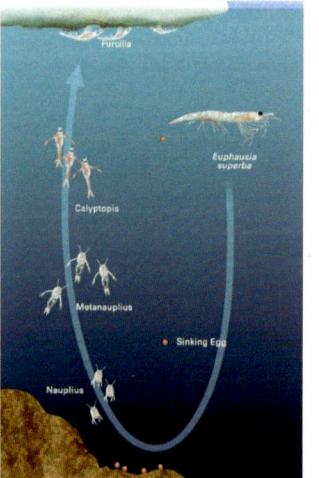

included in the models alongside seal, whale and penguin population health and numbers and a suite of environmental parameters that drive the ecosystem.

One of the major environmental drivers in the Antarctic and Southern Ocean is sea ice extent and duration. Krill are highly reliant on sea ice for their reproduction and growth. Juvenile krill use the sea ice as habitat that they can hide from predators in. They also eat the abundant sea ice algae growing on its underside during their 1st winter, a time of little food in the water. Krill reproduction is even timed. Females spawn their eggs in the late summer/early autumn to descend to the sea floor. It takes nearly 3 months for juvenile krill to rise through the water column as they undergo several metamorphoses. Juvenile krill will reach the surface in their furcillia stage reach the surface in winter, when the sea ice is at the greatest extent. Sea ice is so interrelated with krill juveniles surviving their 1st winter that krill populations follow the same trend as sea ice, with years of larger sea ice having larger numbers of krill surviving into the population the next year and vice versa.

With decreasing levels of sea ice and shorter durations of sea ice in the winter due to climate change, krill populations are expected to decrease into the future. However, not all of Antarctic's sea ice is shrinking at the same rate. In the South Atlantic, sea ice is decreasing rapidly but in the Pacific and Indian Oceans, sea ice and temperatures are remaining far more stable. This may mean that the South Atlantic may no longer be the area of greener pastures it once was for the krill fishery and that the Indian and Pacific Oceans may become more pivotal areas for krill recruitment and reproduction.

Currently CCAMLR only allows 1 krill harvesting vessel in the Indian Ocean (from China) and it has only been doing exploratory fishing there since 2017, whilst the entire Pacific Ocean is a no harvesting zone (except for scientific purposes). Due to the strict geographical management of where krill can be caught and how much can be caught in each area by CCAMLR, there is significant hope that krill will be able to be sustainably managed for many years to come.

A poem of Antarctica memories
by Sarah Charnaud, Ph.D.

The smell of a crevasse - the old and fresh smell trapped in the blue cavern
in the snow;

The sight of green when we returned! It filled my eyeballs with life and
made me appreciate our daily environments just as much.

The surprising swathes of the green mosses at Bailey Head.

The tabular icebergs — thousands of years of ice, they could have been
from Larsen B, cleaved off in 2002 and slowly and inexorably breaking up,
melting, rising the ocean.

But for now, when they roll and show their lumpy underbellies providing a
sailing spot for penguins.

We saw these 7th December with the minke whales breaching and
corkscrewing through the water.

Sarah Charnaud, Ph.D.
Malaria Researcher, Walter and Eliza Hall Institute
Victoria, Australia (originally from South Africa)
Photo credit: Shelley Ball

Carol Devine
Médecins Sans Frontières/Doctors Without Borders
Toronto, Canada
Photo credit: Carol Devine

Deborah Pardo, Ph.D.
At this time: Population Modeller, British Antarctic Survey
England, United Kingdom (originally from France)
Freelance scientist, Marseille, France.
Photo credit: Deborah Pardo

Joanna Young
Ph.D. Candidate; Co-founder of Girls on Ice Alaska, a program of Inspiring Girls Expeditions
Alaska, United States
Photo credit: Joanna Young

Anne Christianson, Ph.D.
Pew Charitable Trusts, Senior Associate, Protecting Antarctica's Southern Ocean
University of Minnesota, Natural Resources Science and Management
Minnesota, United States
Photo credit: Sarah Conolly

HOMEWARD BOUND FACULTY AND CREW

Fabian Dattner
Homeward Bound Founder
Photo: Olie Sansom

A note
by Fabian Dattner

Every face you've seen, every story you've read, every picture you've looked at and wondered about was shared and lived by the first cohort of Homeward Bound, 2016.

These were the women who had the personal courage to join an initiative that was absolutely in its infancy in early 2015. They understood in their bones this call to action to women with a STEMM (Science, Technology, Engineering, Mathematics, Medicine) background was essential to our planet's future. These were the women who felt they had a contribution to make, trusted the vision, and stepped in without precedent.

Perhaps we all entered into the inaugural Homeward Bound experience with sparkles in our eyes. Most memorable for me was the moment I came down from my room into the foyer of the Ushuaia Hotel, on the evening we all met for the first time. Everyone was waiting to go to dinner, and we had some distance to walk. The excitement, joy, and hope rising up to greet me was jaw-dropping.

The whole trip was an amazing experience, and this book is a testament to where we went and what we did.

The program taught me Homeward Bound's success was going to require more courage from me to step up to meet the expectations of these brilliant women. It also needed significant changes to the program including greater awareness of the audience and their challenges, a deeper focus on the gender challenges in STEMM and far wiser management of the experience, for the faculty as well as the participants.

We received significant feedback and, to the best of our ability (it wasn't always easy) we implemented what was asked for. In fact, we learned so much from this process, we have made it a feature of every Homeward Bound program (formally reviewed by Dr. Megan Oaten, Griffith University alum HB 2 & 3).

Today, we have three successful programs under our belt, the inclusion of some 48 nations and 36 sciences in HB4 and hundreds of Homeward Bound participants in collaborations, getting promotions, doing TED talks, influencing policy and decision making. We could say we are coming of age, but we don't. We still try as hard, listen as intently.

Today, many alumni lead various aspects of Homeward Bound. My co-founders (Dr. Justine Shaw & Dr. Mary-Anne Lea) together with core faculty, Business Hub, and Board are inspired by where we are and deeply committed to where we could be.

Our planet is in crisis, and the practice of leadership globally is struggling. Homeward Bound is a transformational leadership program for women with a STEMM background. We have an estimated global audience of some 1.2 billion people (March 2019) and many thousands directly linked through social media and our newsletter. A film coming out at the end of 2019 is for everyone involved with Homeward Bound and for those who believe we could be managing our planet with a focus on collaboration, inclusion and a legacy mindset that includes being trusted with assets, both money and people.

Homeward Bound has a target of 1,000 women participants by 2026.

We hold fury and hope in our left and right hands.

We belong to a whole whose parts are greater than any individual.

Fabian Dattner
Dreamer, Co-Founder, CEO

June 2019

Kit Jackson
Homeward Bound faculty

Songqiao Yao
Homeward Bound faculty

Mary-Anne Lea
Homeward Bound faculty

Justine Shaw
Homeward Bound faculty

Monika Schillat
Expedition Leader for Antarpply Expeditions, lead author of Tourism in Antarctica:
A multidisiciplinary view of the new activities carried out on the white continent

Julia May
Co-founder, Visibility stream, Homeward Bound
Director, Visibility Co
Photo credit: Julia May

Greg Mortimer
Expedition Leader and Eminent Moutaineer

Captain Waldemar Wichmann
Antarpply Expeditions

Dilia Martínez Méndez, MD
Senior Medical Officer and Antarpply Doctor
Photo credit: Natasha Mier

Marshall Cowley
Homeward Bound Design Architect
Photo credit: Marshall Cowley

Before we go, I want to add one more essay. Jessica Shamek is a Ph.D. student at Portland State University. She did not participate in the 2016 Homeward Bound excursion. But I want to include her essay for the up and coming women scientists who work hard and have ambition. We are not going to sugarcoat it. Science is hard and society is harder. This essay is dedicated to future Antarctican scientists. The ones with grit. The ones who are going to change the world.

Jessica Shamek
Ph.D. Student Eppley/Rosenstiel Lab, Department of Biology
Portland State University
Oregon, United States
Photo credit: Jessica Shamek

If I go to Antarctica...

By Jessica Shamek

At 7 years old, I developed a healthy obsession with dinosaurs and declared my love of paleontology but was told by my aunt that I wouldn't make any money and I should go into business. At 14, I developed a fascination with Einstein and his theory of relativity and then at 17, I learned about Schrodinger's theory. It never crossed my mind that I might become a scientist someday, or that I might go to Antarctica for that matter. My love of science was there, but I didn't see it for what it was until my early 20's.

My experience is not unique, in part because our society has historically excluded womxn from participating in academia. This continues to manifest in the present day, in the form of missed opportunities to foster the love of science, safety, and success of womxn. We need to work towards the inclusion of womxn in science at every level. A scientific culture that is friendly to womxn of all races and background, one that views womxn as necessary contributors to the advancement of scientific knowledge.

I eventually began to fully explore my love of science by working at plant nurseries and then returning to school. My first paid field position as a plant ecologist was an eye opening experience. I saw first-hand the lack of support and safety offered to womxn who do field science. I grappled with comments from coworkers and supervisors about my body and my politics - two things that need not be discussed at work. I was cat called inside of a federal office. I struggled with the double standard my work was held to when compared to my male colleague and fought harder than I should have to prove my skills and abilities. The season ended with me filing a formal complaint against my crew lead and giving my supervisors a literature review I had compiled on best practices for preventing gender discrimination in field science. I deserve so much more than that. All womxn do.

If I go to Antarctica, will I have to face that - or worse - all over again? In the most remote place on Earth? In a 2013 survey of academic field experiences conducted by researchers at the University of Illinois and published in PLOS One, 72.4% of respondents (both men and womxn) reported they had directly observed or been told about the occurrence of other field site researchers and/or colleagues making inappropriate or sexual remarks at their most recent or notable field site. A further 64% of both men and womxn reported that they had personally experienced sexual harassment in the field. Within this group, gender was significant in predicting if a person experienced sexual harassment in the field, with 3.5 times more womxn than men experiencing it (Clancy et al. 2014).

Knowing that gives me pause. I should be thrilled to be on my labs list to go to Antarctica, as a first generation womxn in science, a curious person, and a first year Ph.D. student. Getting picked to do research in Antarctica for my lab could be an incredible experience. I'm a scientist who could go to Antarctica. That's so cool. And if I go? Maybe I could go to Antarctica.

I imagine the sharp breath taking edges of the glaciers that meet the sea and what the cold air might feel like against my cheeks and throat. Most people probably imagine Antarctica as an untouched wilderness of snow and ice. But when I think of it, as a scientist, I imagine the algae and fungal spores that thousands of researchers have brought in on their clothes and feet. I imagine the carpets of moss, lichens, and liverworts, the organisms that I study, growing in vast carpets on what at first appears to be a barren landscape. There are approximately 130 species of bryophytes and 200 species of lichen on the continent compared to the 2 species of vascular plants. It's a harsh environment, no question, but life survives. It finds a way to thrive. Like I did, despite sometimes being actively discouraged from studying science.

I'm still full of questions. My imagination runs rampant on every variable. Would it be worth the risk? What kind of science could I do in Antarctica? How valuable would it be? Do I even want to go, knowing the hardship that may lay ahead? I sit at my warm kitchen table in pajamas typing this and pondering these questions while I recover from my first term of graduate school. What a wild ride, life is. What an interesting place our planet. I wish it was a better, safer place for everyone. Until then, I'll imagine what that safer world will look like and strive to make it happen.

Note: This essay uses the word "womxn" as a new and inclusive orthography for the word "women", as it is indicative of the prejudice and institutional barriers that all womxn - trans womxn, womxn of color, and non-binary people - continue to overcome.

"Imagination is more important than knowledge. For knowledge is limited to all we now know and understand, while imagination embraces the entire world, and all there ever will be to know and understand."

-Albert Einstein

CONCLUSION

I put together this book to showcase women in science to offer another side of the male-dominant story about Antarctic research. Social scientists use the term "multiple truths" to express the idea that there is not one "true" reality to be discovered, but there are as many different ways of interpreting the world as there are humans. Some argue that nature, like Antarctica, has a 'truth' of its own. After understanding this core concept, we can take the next step of critical inquiry where we can ask whose truth dominates, when, and how? Does a Western, masculine culture dominate science? When push comes to shove, do humans put ourselves first, or do we consider what is sustainable for the planet? As a society, we should start having these conversations.

This book is an "unauthorized" account, meaning that I'm doing this on my own accord. The views and opinions do not reflect Homeward Bound, and some of the women here might agree or disagree with certain aspects. The book was nearly three years in the making. After returning from the trip, I contacted each person individually for their preferred photo, name and title, and even traveled to Australia to follow up with some. As you can see, some scientists went above and beyond by contributing chapters, interviews, and quotes. All but one participant agreed to be included. I shared numerous drafts with faculty, staff, and participants so that any one at any time could comment about the contents of the book. I prioritized collaboration. I had to give up a little control and take a little more time to do it, but I think it accurately captures the sense of becoming in a place like Antarctica.

I hope from reading this book, you get the sense of all the hard work the faculty and crew put in before, during, and after the expedition. I would like to extend my warmest gratitude to these beautiful and amazing people. Homeward Bound plans on conducting a total of ten visits to Antarctica. If you are interested in applying for a future trip, see:

https://homewardboundprojects.com.au/.

I titled this book "Antarctica Becomes Her" for a few reasons. First, it is important for young girls to see women scientists in Antarctica in a positive way. For too long, the discipline of science has been seen as "unbecoming" for women to undertake. But women too have deep curiosities about how the world works, indeed, it is a uniquely human characteristic. Science is simply the method of formulating research questions and testing hypotheses through systematic experimentation. That's it. You don't need to be a certain gender, or race, or class to do it. You don't even need to wear a lab coat or peer through a microscope. The exciting part is not finding the answer but finding new questions to ask. When I say "Antarctica Becomes Her," I mean Antarctica is a place of scientific exploration, and as humans travel there to study the continent and the role it plays in our changing climate, I want more women to be a part of that adventure. "Becoming" is an adjective. Antarctica looks good on you!

I also use "becoming" as a noun, the process of growth and development as we passed into another state of understanding. The Homeward Bound experience was transformative for everyone involved. I don't think any one of us could say that we came back as the same person.

I use "becoming" as a verb. We "became" a part of Antarctica. The thing that makes adventure so adventurous is the point from which you go from not knowing to knowing. There is nothing more attractive than a question mark.

Just as Antarctica is now forever a part of us, we will always be a part of Antarctica. And all of the future female explorers who will travel to Antarctica and all the people of color and people with differing abilities and those of all nations who will tell the story of what they see and experience when they go to Antarctica. We have a limited view of Antarctica when perspectives are constrained to the elite, dominant classes. And that is why, although the impact of human visitation to the continent is so detrimental, I cannot do anything but encourage everyone to go, to become Antarctica and to become themselves.

About the Author

Melissa Haeffner, Ph.D., is an Assistant Professor in the Environmental Science and Management department at Portland State University. Her research unifies several domains that contribute to the knowledge of local politics in watersheds and how they shape urban water infrastructure development in the past, the present, and under future predictions.

Her ongoing research and teaching commitments investigate water insecurity and justice within municipal water systems and the links between multi-scale policies, infrastructural and environmental conditions, and household behavior.

Her work focuses on "just water" and how social, political, and biophysical factors structure access to water, using the concept of environmental justice to draw attention to issues of fairness and equality in the ways different social groups gain access to natural resources. Dr. Haeffner received a BA/MA in Sociology from DePaul University and a Masters in Science through the Department of Urban Studies and Planning from the Massachusetts Institute of Technology. She received her Ph.D. from Colorado State University (CSU) in the Graduate Degree Program of Ecology, with a specialization in human-environment interactions. She is the author of the book Water Walkers: portraits of Ghana's street vendors, based on qualitative fieldwork, and several scientific papers.

Acknowledgements

Thank you to the faculty, crew, and participants of the 2016 Homeward Bound trip. Thank you to everyone who supported us financially, emotionally, and intellectually. I want to say a special thank you to my parents, Walt and Vickie Haeffner, who have always supported my education, international travels, and crazy ideas.

Citations used in the book:

Blackadder, J. (2015). Frozen voices: Women, Silence and Antarctica. Antarctica: Music, Sounds and Cultural Connections, 169-177.

Clancy, K.B., Nelson, R.G., Rutherford, J.N., & Hinde, K. (2014). Survey of academic field experiences (SAFE): Trainees report harrassment and assault. PLoS One, 9(7), e102172.

Editor in Chief:
Melissa Haeffner

Production Manager:
Jay Nadeau

Portraits:
Melissa Haeffner, unless otherwise noted

Other photography credits:
© Carol Devine, page 138
© Shelley Ball, page 145
© Carol Devine, page 146
© Deborah Pardo, page 147
© Joanna Young, page 148
© Sarah Conolly, page 149
© Olie Sansom, page 151
© Julia May, page 159
© Natasha Mier, page 162
© Marshall Cowley, page 163
© Jessica Shamek, page 164

Figures
© p 141-143 (Nicole Hellessey)

Designer:
Stevon Christopher Burrell
https://www.linkedin.com/in/burrellstc

Layout:
Dena MT Eaton
denameteaton.com
Marketing Coordination:
Jessica Shamek
Sirajum Munira
https://www.behance.net/sirajummunira

Research:
Greg Mortimer
Wynat Smith
Ida Kubiszewski
Logan Christianson
Amy Vale

Proofreading:
Sarah Chaurnaud
Fern Hames
David Perry